The Spirit of the Storm:

A Collection of Interviews from the Gulf War Era

AUTHORS

MAJ Thomas Bramanti, US Army
Maj Erika Chute, US Air Force
Maj Heather Demis, US Air Force
LCDR R. Thomas Griffin, US Navy
Maj Stanley Grodrian, US Air Force
Maj Jessica Guarini, US Air Force
Maj Tenaya Humphrey, US Air Force
Maj Christopher Jefferson, US Air Force
Maj Fred Merrill, US Air Force
Maj Shannon O'Neal, US Air Force
Maj Will Powell, US Air Force
Maj Michael Ryan, US Air Force
Maj Adrienne Stahl, US Air Force
Maj David Underwood, US Air Force
Maj Dustin Wade, US Air Force
Maj Vanessa Wilcox, US Air Force
Chaplain (Maj) Christian Williams, US Air Force

EDITORS

Dr. Lisa Beckenbaugh
Maj Tenaya Humphrey, US Air Force
Dr. Jared Donnelly

Library of Congress Cataloging-in-Publication Data

Gulf War: Spirit of the Storm / Authors; Editors.

ISBN: 978-0-692-84665-0
1. Gulf war, 1991 — Personal narratives, America. 2. Soldiers – United States – Interviews.
3. Leadership – Lessons, Personal. 4. Oral History – Military, Airpower.

For more information about this publication and additional copies contact:
Gathering of Eagles Foundation
PO Box 1062
Montgomery, Alabama 36101-1062
Email: acscgoe@goefoundation.org
www.goefoundation.org

All of the proceeds from the sale of this book benefit the Gathering of Eagles Foundation.

To the Air Command and Staff College Class of 2017, this book is for you.

To the Eagles, past, present, and future, thank you for letting us share your stories. May your legacies of courage, honor, and sacrifice be passed down through the generations of all those who follow in service to country.

EYRIE

Aviation legends are replete
With stories of some men who meet
In Hallowed halls,
And there they're joined by younger men
Who'll man the planes in defense when
Our Country calls!

A thousand deeds recited there,
Tales of daring in the air
By warriors bold,
Who met the foe for valiantly,
And fought with highest gallantry
Like knights of old!

When searing winds of war would scorch
The earth and freedom's holy torch
Was dimly lit,
With the strength that courage brings,
They formed a canopy of wings,
And sheltered it!

From generation to generation,
The centurions of this nation
Have passed along
The values cherished by each age
Like Duty, and Country, a heritage
That makes us strong!

Now brave young eagles proudly stand
And accept from aging hand
That legacy,
Sentinels who in wondrous flight
Are guardians of the sacred light
Of liberty!

So of all the place that I know,
If I can chose one place to go
I would rather
Visit that special eyrie
At a time when I can see
Those eagles gather!

■ Captain Harry W. Brown
WWII Ace and 1989 Eagle

FOREWORD

Since 1982, the Gathering of Eagles program has included combat leaders from WWI, WWII, Korea, and Vietnam, the wars in the Middle East, plus many civilian luminaries. This year, the program is focused exclusively on the Gulf War of 1990-1991, DESERT SHIELD (the buildup) and DESERT STORM (the hostilities).

DESERT STORM was a defining time in America's military history. It was the first time advancements in GPS, stealth, and precision guided munitions were integrated into the battlespace, which gave us an overwhelming technological advantage. The military might unleashed on Iraqi Armed Forces resulted in an astounding victory for US-led Coalition Forces.

The following collection of interviews captures the experiences, the spirit, the wisdom, and the heroism of those who not only shaped, but were shaped by, the Gulf War. Some experiences were of the theorists, strategic thinkers, and planners who constructed the overall vision for the operation and brought it into the field. Other experiences were those at the operator level, directly prosecuting the mission at great personal risk, including the awful experience of being a Prisoner of War.

Each of these Eagles, honored by Air Command and Staff College (ACSC) for their contributions to air, space, and cyberspace, provide stories from their formative years, leadership lessons, and a unique perspective on the Gulf War. Together, these interviews paint a picture of the era and broaden our collective understanding of the full landscape of DESERT STORM and its human dimension.

The authors of this work, Majors at ACSC on the cusp of squadron command and staff assignments, felt it important to keep the words and thoughts of the Eagles organic, even raw, resulting in the following interview-style manuscript. In addition, the authors provide an account of their personal experiences with the Eagles, as well as lessons in leadership they gleaned from the process. The result is candid, reflective, and provocative, with uncensored views, philosophies, and opinions from our legacies—these great practitioners of airpower.

As a former Eagle, and Air University Commander during the first Gathering of Eagles (GOE) event, it makes me extraordinarily proud of all those who have kept this tribute to aviation heritage alive for the last 35 years. I extend my sincere thanks to all those ACSC students and faculty members who have contributed to ensuring that the legacy of the Eagles has continued for nearly four decades. I hope this book helps to forge the spirit of service to our Nation within each of you, or at the very least, helps you appreciate that service.

■ Lt Gen (ret) Charles "Chick" Cleveland
Former Air University Commander 1992 Eagle

ACKNOWLEDGEMENTS

First and foremost we would like to thank Dr. William Dulaney, who encouraged us to write this book, despite our reservations about finishing it during our year at ACSC. Dr. Dulaney, you have the heart of a warrior, an artist, and a true teacher. Thank you for making us believers. Get well and come back soon!

To our editors and advisors, Dr. Lisa Beckenbaugh, Maj Tenaya Humphrey, Dr. Jared Donnelly, Maj Johnathan Artis, and Maj Jonathan Stinson, thank you for shepherding us through this process and for scouring every word of this book. To our copy editor, Rebecca Donnelly, thank you for your comprehensive review and attention to detail. Your collective efforts turned our individual chapters into a book that honors our Eagles and preserves their legacies.

To the ACSC leadership, Col Brian Hastings, Dr. Jim Forsyth, Col Scott Yancy, Col Leon Perkowski, and Col Richard Cooney: thank you for your encouragement, enthusiasm, and for allowing the students to take ownership of this program. Thank you to all the ACSC faculty members who supported the GOE team this year and who contributed to the writing of this book with brilliant ideas, thoughtful edits, and moral support. A special thank you goes to Dr. Kevin Holzimmer, Dr. Kelly Grieco, Dr. John T. LaSaine, Dr. John Terino, and Wg Cdr Rich Milburn, RAF for their help.

Thank you to Col (ret) Larry Carter for sharing his memories of the 1982 "Great Moments in Aviation History" event and his experience of escorting Gen (ret) Paul Tibbets. These memories brought the GOE history to life.

To AEgis Technologies, thank you for graciously donating your time and energy for our graphics development effort, with a special note of appreciation to Steven Hill, William Watson, and Robert Gravitz, who made this effort possible. To Grits Morris at Walker360, your gift for formatting helped to make this book better than we ever thought possible.

Above all else, we want to thank our eight Eagles. They live continuing lives of service and dedication to country. They have set the example for all to emulate and are solidified in the history of our armed services. What we do today and in the future would not be possible without your sacrifice. We thank you for everything you have done, and for sharing your stories with us

TABLE OF CONTENTS

DESERT SHIELD/DESERT STORM

Area of Operations: 1990-1991

Source: *War in the Persian Gulf Operations Desert Shield and Desert Storm, August 1990–March 1991,* Center of Military History, United States Army, Washington, D.C., 2010.

INTRODUCTION

At the ACSC Graduation Ball on 6 June 1982, comedian Bob Hope surprised the class by delivering an opening monologue for the first GOE program, known at that time as "Great Moments in Aviation History." In keeping with Bob Hope's sense of humor, he told the class he had "just heard about the event and had decided to show up." In reality, it was the Air University Commander, Lieutenant General Charles "Chick" Cleveland, who had coordinated for Bob Hope's appearance. This event was two years in the making and was inspired by General Paul W. Tibbets' visit to ACSC in 1980. Gen Tibbets had shared his experience piloting the Enola Gay, the B-29 that dropped the first atomic weapon on the Japanese city of Hiroshima. He told the class that he knew the bomb had detonated because he could taste the lead in the fillings of his teeth. He went on to tell the students about meeting Japanese Navy Captain, Mitsuo Fuchida, the pilot who had led the first wave of air attacks on Pearl Harbor. General Tibbets had met Capt Fuchida at a conference in 1958 where the Captain had told him that his actions had undoubtedly saved millions of lives during WWII. Capt Fuchida believed that had the Emperor of Japan not been persuaded to end the war, every Japanese man, women, and child would have defended him to the death. ACSC leadership understood the profound impact of Gen Tibbets' stories on the class and encouraged the faculty to develop an aviation heritage program to further develop the study of aviation history and the contributions of air and space pioneers.[1]

The inaugural GOE team selected fifteen distinguished aviators and invited them to share their unique personal experiences through a series of interviews and social events with members of the ACSC class. This group included military and civilian aviators who had served in the Air Force, Navy, Marine Corps, Army, and National Aeronautics and Space Administration (NASA). The first Eagles as they would later be called, spanned aviation history from WWI to the Space Age.

The famed 1982 Eagles were:

- Neil A. Armstrong, who, on 20 July 1969 became the first person to step onto the moon proclaiming, "That's one small step for man, one giant leap for mankind."
- General James H. "Jimmy" Doolittle, famed air racer and instrument flying pioneer, who led the first retaliatory strike against the Japanese mainland following Pearl Harbor.
- Brigadier General Joseph J. Foss, "The Flying Marine," Medal of Honor recipient, and the first American to tie Eddie Rickenbacker's American WWI record of 26 enemy kills.
- Colonel Francis S. "Gabby" Gabreski, then America's top-scoring living ace with 28 victories in the WWII European Theater of Operations and 6 ½ confirmed kills in Korea.
- Lieutenant Commander George Henry Gay Jr., sole survivor in his flight of 15 torpedo bombers, launched from the Hornet aircraft carrier during the Battle of Midway.
- Colonel Gail S. Halvorsen, "The Candy Bomber," who conceived and founded "Operation Little Vittles," dropping candy by miniature parachutes to children during the 1948 Berlin Airlift.
- Colonel William J. "Pete" Knight, who holds the world speed record for winged powered aircraft. He flew Mach 6.72 or 4,534 miles per hour in his C-15A-2.

1 Col (ret) Larry Carter, interviewed by Maj Tenaya Humphrey and Maj Heather Demis, 24 March 2017.

- General Curtis E. LeMay, one of the most influential, operational air commanders in WWII and the driving force behind Strategic Air Command in the Cold War-era Air Force.
- Colonel John W. Mitchell, P-38 pilot, who planned and led one of the longest fighter interceptor missions of WWII to shoot down Japanese Admiral Isoroku Yamamoto.
- Chief Warrant Officer Michael J. Novosel, medevac helicopter pilot and Medal of Honor recipient, who performed 15 extremely hazardous extractions and rescued 5,500 soldiers during Vietnam.
- General Robin Olds, legendary combat commander and ace with 13 victories in WWII plus four in Vietnam, who led F-4 (Phantom IIs) in the famous fighter sweeps over the Red River Valley in North Vietnam.
- General Paul W. Tibbets, pilot of the Enola Gay, the B-29 that dropped the first atomic bomb in WWII on Hiroshima, Japan, which changed the world and ushered in the nuclear age.
- Lieutenant Colonel George A. Vaughn Jr, the United States' fourth-ranking WWI ace with 13 aerial victories in the SE-5 and the Sopwith Camel.
- Major General Leigh Wade, pilot of the Douglas World Cruiser "Boston" who accomplished the world's first aerial circumnavigation of the globe in 1924—three years before Lindbergh flew from New York to Paris.
- Brigadier General Charles E. "Chuck" Yeager, double ace with 13 enemy kills, first man to fly faster than the speed of sound, and pilot of the Bell X-1A, which flew 2 ½ times the speed of sound.

The first gathering in 1982 made history and cemented the GOE tradition of annually honoring a select group of aviation heroes and pioneers as the capstone event of ACSC. Future Eagles would include Medal of Honor recipients, Tuskegee Airmen, Doolittle Raiders, Women Air Force Service Pilots, Astronauts, a United States President, and many others who made significant contributions to air and space. Through the years, countless ACSC students, faculty advisors, and community leaders have dedicated their time and talents to the GOE program, enabling the mission of bringing distinguished Eagles from all over the world to inspire future generations. This year, a staff of sixteen students selected eight Eagles from the Gulf War Era. In teams of two, students traveled to interview each Eagle and record their stories. By capturing the words of the Eagles in this book, the GOE experience can be shared with not only ACSC but with airpower and military history enthusiasts all over the world. In keeping with the GOE tradition, the eight Eagles will travel to Maxwell Air Force Base (AFB) in May 2017 for a week of events to honor their contributions. During this time, the Eagles will participate in live interviews with an audience of civilian and military personnel from ACSC, Maxwell AFB, and members of the surrounding community. The 2017 GOE mission statement "Honoring our past to inspire military and community leaders in a new era" drives these events and provides the motivation for this book.

While in recent years GOE has focused on honoring legacies from the WWII era, the 36th GOE event will jump ahead half a century to honor Eagles from the Gulf War era. The title of this book, *The Spirit of the Storm: A Collection of Interviews from the Gulf War Era*, seeks to capture the human spirit behind the operation. It includes the experiences that shaped the Eagles leading up to DESERT STORM, their personal tragedies and triumphs during the operation, and the impact of this time on their personal and professional lives. The following is an excerpt from a speech United States Air Force Chief of Staff, General David Goldfein gave at the 2016 Air Force Association Air, Space, and Cyberspace Conference. Gen Goldfein explains his own formative experiences in the Air Force, why it is necessary to examine events preceding DESERT STORM, as well as the force that evolved after Vietnam.

You know, you can't have a discussion about DESERT STORM without first talking about Vietnam, because the reality is we came out of Vietnam, and a certain number of airmen stayed with the company, and they rebuilt our Air Force, and they took us from the days of the F-4 into the days of the F-15 and the F-16, and they fought to bring this all-volunteer force, this very professional force, into being.

It was those airmen that led that effort, not the least of which are our former Chiefs who are in the front row, and many of you who are here in the room, that allowed us to have the force that we went into for DESERT STORM.

Let's talk a little bit about DESERT STORM and my experience. As a young flight commander at Shaw Air Force Base, we got the call shortly after Saddam Hussein rolled south, and 48 hours later, we were rolling forward, and we were launching our 17-hour sorties into the Middle East. I got a chance as a young captain to watch an entire wing mobilize for warfare, and everybody put their shoulder against the wheel, and everybody pushed forward to make it happen.

Whether you want to talk about the services folks who got the meal ready in the DFAC or the munitions loaders and builders who actually built up the munitions and made sure we had what we needed, because at the time we didn't know whether we were going to fight our way into enemy territory, especially as we crossed north of Libya.

If you want to talk about the air traffic controllers who put together all the flight plans and got us off the ground in a completely [communications out] environment, so we could mask our deployment, as though 24 F-16s taking off is going to mask anything, but we thought it was masking us. To the maintainers across the line both in the back shops, they got our aircraft ready and got us airborne.

What I learned as a young captain and what I came to believe from that experience is there were literally hundreds of fingerprints on my aircraft that made that deployment happen. I came to believe that every airman matters, everyone is required to make the mission happen.

As I taxi out to launch into the great unknown, as many of us did, lining the taxiway were all those airmen, active, Guard, Reserve, civilian, industry partners with their logos. As we all taxied by, every one of them gave a little salute, a thumbs up, a wave. They were all there for us.

What you need to know about me first and foremost is that I truly believe every airman matters.

When we got into theater, and we were facing the great unknown, it was interesting for us because we had one combat veteran in the entire squadron, it was our squadron commander, at the time, Colonel Billy Deal. It actually is true that you can mission plan in whatever time is left to you, because we mission planned one mission for seven months, and as the mission planner I remember working with General North, who I know is here, and he would take off to Riyadh, and he would take some kind of contraband with him, I'm not sure what, and he would come back with stacks of maps, and we would plan, we would plan, rehearse, we would train, and we would plan again, and we would go out and rehearse, and we would have these conferences and talk to our joint partners. When it came time for the actual mission, none of us knew how we were going to do. We were all wondering, "What is this combat thing going to be all about. Are we ready? Are we trained?" That pilot's prayer that every combat veteran whether it is a pilot or anybody else always prays, which is, "please, God, don't let me let my buddies down, I hope I'm ready for this," and we take off and we launched the first mission, and we get across the line and a voice, I'm not exactly sure where it came from, some voice actually came on the radio and said the Iraqi military has been informed you have crossed the border.

I remember thinking at the time, "well, hell, now I'm just going to die tense." In actuality, we are in the great unknown, we are in our first combat, we look out, and here we have a commander from Vietnam, Col Deal, and he's calling out things to us like it's a walk in the park, and he says, hey, there's AAA, anti-aircraft artillery at two o'clock. We all looked at it. He says there is a surface-to-air missile, left ten [o'clock], sure enough, we saw surface-to-air missiles, and we saw them exploding above us, and we all stared at them. Then we hear "Splash Mig-29," and we see an aircraft hit the ground and explode. I will tell you I remember that moment in my cockpit. You can't imagine the calm that came over my cockpit when I recognized and realized that I was in a very familiar environment. I had heard all these radio calls. I had actually seen AAA and SAMs smoke. I hadn't seen an airplane hit the ground yet, but everything else, the formation, the radio calls, the environment, the intensity, the adrenaline, I had seen it all at Nellis Air Force Base, at RED FLAG.

What I realized in my cockpit was, "okay, I can do this." We went in and we crushed the target. When I came back and talked to all my flight mates afterwards, I said, "what did you think?" They said, "Man, just like RED FLAG." For those who built that exercise, to give young aviators the first 10 sorties so they could survive combat, gentlemen and ladies, it worked. I came to believe that high-end training against the most difficult threat in the most difficult environment is nothing short of a moral obligation, and we had proved it in DESERT STORM. It wasn't that many years later when we saw ethnic cleansing again, and this time it was in Kosovo, in a neighboring country. But now, I'm a squadron commander. Now you have a different question. When you're a young captain, you are wondering, "Am I ready? Am I trained? Am I going to let my buddies down? Am I going to be able to do the mission that I've been trained to do? Am I ready for this?" As a squadron commander, you have a little bit different question. You start asking, "Have I trained them enough? Was I hard enough on them? Did I give them the skills they need? Am I ready to send them into this conflict? Am I ready to lose one of them?"

I remember Gen Deal or Col Deal told us one time during DESERT STORM; he looked at us, he goes, "You know, you're like my kids." He said, "You're not the best I've seen, you're not the worse I've seen, but you are all I've got." You know, you feel like that as a squadron commander. You know what? They do great. As I led them into that first night's sortie and then back, I got them all in a group, and we were debriefing, and I said, "Okay, what do you think? How did you do? What did you think?" They all looked at me and said, "Sir, it was just like RED FLAG." History may not repeat, but it sure rhymes a lot.[2]

As Gen Goldfein articulates, the events of DESERT STORM cannot be fully understood without considering the events that both preceded and followed the conflict. Just as the leaders of DESERT STORM were the lieutenants and captains of

2 Gen David Goldfein, United Stated Air Force Chief of Staff (address, Air Force Association Air, Space and Cyberspace Conference, Gaylord, National Harbor, Maryland, 20 September 2016).

Vietnam, the lieutenants and captains of DESERT STORM would go on to lead the force we have today. Their experiences shaped the way these airmen led and helped determine the course the Air Force would take in the aftermath of each conflict. While most historical accounts of DESERT STORM relate what happened, this collection of interviews seeks to answer the why and the how questions. Our aim is to capture what other accounts may have missed, in order to create a more complete picture of the conflict, and what it felt like to serve and lead during DESERT STORM. We want to view this time through the eyes of those who served, and read the history in the words of those who lived it. We want to know if DESERT STORM fulfilled the promise of early airpower theorists or if it was simply the "right" war at the "right" time. We ask if DESERT STORM is a model that one should look to in preparing for future conflicts: is this "symmetrical" force-on-force engagement of continuing relevance for the future development of the joint force? We seek to understand the service culture of the 1980s and 1990s to determine what has changed and what has stayed the same. We want to look specifically at leadership and ask, "What makes a good leader? Are there traits that all leaders possess? In what ways is leadership contextual? Are there lessons that we can take from those who led in DESERT STORM which apply today?" It is with these questions and more that we embarked on this project.

This book is divided into two parts in keeping with the 2017 GOE theme: Honoring Legacies, Forging Leaders. Part I is dedicated to honoring the legacies of the Eagles. Each of the eight chapters in this section is devoted to an Eagle and includes an edited transcript of the interview conducted by GOE team members. The format is question and answer to allow the Eagles to tell their own stories. Also included in each chapter is supplementary information to provide background information and context. Following the eight interviews is Part II, which focuses on forging leaders. In this section, each team describes the experience of interviewing an airpower legend and offers a leadership lesson they took from this experience. The study of leadership is foundational to the ACSC curriculum and provides an opportunity for students to reflect on their own leadership attributes. Each reader will internalize something different from reading the interviews with the Eagles, and Part II demonstrates the process of deriving personal lessons by studying the words of great leaders. We hope this work provides an opportunity for each individual to glean lessons vital for their further growth as leaders.

Gen Goldfein reminded us of the saying, "While history may not repeat itself, it often rhymes." We have found this to be true as we listened to the words of those who served before us and hope that these interviews and leadership lessons help illuminate the links between veterans of the Gulf War era and those who serve today, all around the globe. The GOE experience has been a oncein-a-lifetime experience for the GOE team, and we are both grateful and humbled to be able to share the stories of the 2017 Eagles with you. May the words of these Eagles be a source of reflection and inspiration to us all as we continue together in service to our great Nation.

The 2017 Eagles are:

- General (ret) Charles A. "Chuck" Horner, Commander of all US and Allied air assets during Operations DESERT SHIELD and DESERT STORM. He led a devastating air war that enabled the coalition ground forces to achieve their objectives in just 100 hours.
- Major General (ret) Gregory "Beast" Feest, an F-117A Nighthawk pilot, who had the unique experience of dropping the first bombs in two operations: Panama's Operation JUST CAUSE and Operation DESERT STORM.
- Major General (ret) Paul T. "PJ" Johnson, A-10 pilot and Air Force Cross and Superior Service Medal recipient who led a two-ship deep inside enemy territory during DESERT STORM on a search and rescue mission to recover

a downed F-14 crew.

- Chief Warrant Officer (CWO 5) (ret) Guy Hunter, Jr., an OV-10 Bronco Aerial Observer, who was shot down during the opening days of Operation DESERT STORM and spent 47 days in captivity.
- General (ret) Walter Kross, Director of Operations and Logistics for all defense transportation requirements in Operations DESERT SHIELD and DESERT STORM who led the largest transportation operation since WWII.
- adier General (ret) Rhonda L. Cornum, a flight surgeon, who was one of only three survivors of an eight-member UH-60 Black Hawk crew that crashed after taking Iraqi anti-aircraft fire. She was one of only two female POWs in the Gulf War.
- Colonel (ret) John A. Warden III, 20th-century airpower theorist and developer of DESERT STORM's INSTANT THUNDER air campaign. He established the reach-back operation Checkmate in Washington, D.C. area to directly support the war effort.
- Colonel (ret) Thomas E. Griffith, Jr., an F-15E weapons system operator, who was shot down during the early days of Operation DESERT STORM. He and his pilot, Colonel Eberly, evaded for two days until capture by Iraqi forces. He was held as a prisoner of war until his release on 4 March 1991.

Part I:
Honoring Legacies

USAFCENT

Charles A. "Chuck" Horner

Commander, United States Central Command Air Forces

General (ret) Charles A. "Chuck" Horner commanded all US and allied air assets during Operations DESERT SHIELD and DESERT STORM. He led a devastating air war against Iraq that allowed the coalition ground forces to achieve their objectives in just 100 hours of fighting. Horner was born in Davenport, Iowa in 1936, graduated from high school in Des Moines, and enrolled in the University of Iowa in 1954. Even as a young man, Horner wanted to fly jets, and the Reserve Officer Training Corps at the University of Iowa was his ticket to the United States Air Force and flight school. He earned his degree and a commission in 1958 and then completed flight training at Spence AFB, Georgia, and Laredo AFB, Texas. He was awarded pilot wings in late 1959, and after combat crew training in the North American F-100 Super Sabre, joined the 492d Tactical Fighter Squadron in England. Three years later, he transitioned to the Republic F-105 Thunderchief "Thud" and served in the 4th Tactical Fighter Wing at Seymour Johnson AFB, North Carolina. After suffering the loss of his parents, his sister, her husband, and their three children in a tragic car accident, Horner volunteered for combat duty in Southeast Asia. He flew 41 missions in the Thud as a member of the 388th Tactical Fighter Wing at Korat Air Base (AB), Thailand. After six months, he returned to the US and served as an F-105 instructor pilot at Nellis AFB, Nevada. In May 1967, Horner returned to Korat, to again fly the Thud. He flew 70 missions as a "Wild Weasel," attacking enemy surface-to-air missile (SAM) sites. In September, after completing his combat tour, Horner was reassigned to Nellis AFB as an instructor and then liaison officer at the Air Force Tactical Fighter Weapons School. Between October 1969 and August 1975, he served on the staffs of Tactical Air Command Headquarters and Headquarters, United States Air Force, and thereafter attended the National War College. He went on to lead two tactical fighter wings, two air divisions, and the Air Defense Weapons Center. In March 1987, Horner took command of the Ninth Air Force and US Central Command Air Forces. On 3 August 1990, Horner was flying an F-16 to Virginia when he was recalled with the news that Iraq had invaded Kuwait. He accompanied General Norman Schwarzkopf to brief President Bush on the situation and then continued to Saudi Arabia. Horner led all air operations in Operations DESERT SHIELD and DESERT STORM. During Operation DESERT STORM, people around the world came to know Horner and his wry commentaries during daily press briefings. After the Gulf War, he pinned on his fourth star and completed a distinguished career by leading the North American Aerospace Defense Command, United States Space Command, and Air Force Space Command. He retired in 1994 with over 5,300 flying hours and lives near Eglin AFB, Florida with his wife, Mary Jo.

EARLY YEARS

After the Korean War, all male students were required to enroll in ROTC at Iowa. Gen Horner picked Air Force ROTC because "they had the fewest parades." Even though his family had a history of flying, he did not envision being a pilot when he went to college. His cousin, Bill Miles, had been shot down in WWII in 1944. Another friend of his, John Towner, died in a training accident at Luke AFB.

I loved being in the military. I had a lot of fun. Also, see I came from Iowa. What in the hell was I going to do in Iowa? I can't farm. So anything else, my dad was a lawyer, he worked for the VA (Veterans Affairs), and I came from nothing. My dad was in WWI and he went back in the Army in WWII. He was always pissed that I joined the Air Force. He knew I was going to get killed. I had a brother-in-law, my youngest sister's husband was killed. He was a big football star at the University of Iowa and he flew into the ground.

These events neither turned him against nor towards flying; it only helped cement the association of death, heroes, and flying. It took one of his ROTC instructors that taught him how to fly the singleengine, unpressurized, retractable gear, four-seat Ryan Navion, to get him hooked. Gen Horner graduated pilot training in October 1960 amid the drawdown after the Korean War, when there was a surplus of pilots and a shortage of fighter aircraft. Despite this fact, Gen Horner graduated pilot training and started flying the F-100D Super Sabre in the 48th Tactical Fighter Squadron at Lakenheath, England.

PREPARING FOR WAR

There are a few instances that truly change one's outlook on life and if careless, these will pass by without any reflection. That time came for Gen Horner in 1962 while at Wheelus AFB, Libya in North Africa. He was flying on a range as a group of four when his wingman lost orientation and Gen Horner had to maneuver out of his way and stalled his aircraft. He was barely able to recover, only feet above the ground.

You had the incident flying in North Africa as a lieutenant where were you barely able to recover your aircraft that fundamentally changed you. Did that have any effect on you during DESERT STORM?

Oh yeah. It changed my life, and I'm not a zealous person about religion, but I thoroughly believe that God's in charge of my life. So, I'm sitting there, Schwarzkopf and [Secretary of Defense Dick] Cheney get on the airplane to leave me there. I have 27 divisions on the border and two Saudi armored car companies holding the line, and I think, I used the "F" word, and then I said, "I can't do it." Then I realized I don't have to do it. All I had to do is do the best I could. I've had a number of occasions one time, you know, you get ambitious particularly when you're a colonel, you start getting ambitious. I'm a wing commander, which is promotable too, you know, and one of my guys tears the wheel off of an F-15 [Eagle], slides down the runway, screws it up, and a class A accident is death for a wing commander. I'm sitting there feeling sorry for myself; I'll never make general, boo hoo hoo. My wife is the church organist. So we go in early because she has to get her music set up and everything. And there's nobody in the church except she's up in the church organ getting her music set up and everything and I'm sitting there and I open the bible and it says in Psalms, and I'm disappointed because I'm not going to make general and it says, "Disappointment is for the treacherous deceiver." Right there, and I think "Oh Shit! He's nailed me!" And if you think about it, it's absolute truth. If you have faith that God's running your life then what do you have to be disappointed about.

On 16 January 1990 Gen Horner was serving as the Ninth Air Force Commander and dual-hatted as the Joint Forces Air Component Commander (JFACC) for US Central Command (CENTCOM). His boss, General Norman Schwarzkopf, was the CENTCOM Commander, most prominently known for his command over all forces during Operations DESERT SHIELD and DESERT STORM. Gen Horner had taken advantage of a rare break in his schedule to fly his F-16C. Alongside his aide-de-camp, Lieutenant Colonel Jim "Little Grrr" Hartinger, Jr., he had just started what would have been a couple of hours of air combat maneuvers over the Atlantic Ocean against a flight of F-15s. They had just executed a few "G" maneuvers when he received the call he had been anticipating. The Washington Center relayed a message to Gen Horner—return to base and call GEN Schwarzkopf—Secure. The days following the 2 August 1990, Iraq invasion of Kuwait had been a waiting game and he knew a US response was imminent.

What was running through your mind when you were flying and received the call to return to base and call GEN Schwarzkopf?

The problem was that the invasion had occurred and everybody went dead silent; nobody said anything. So I said, "What is going on?" I wasn't surprised when the call came; I was surprised it took them so long to call. I think what happened was, President George Bush was out with British Prime Minister Margaret Thatcher in Aspen, and I think she's the one who kicked him in the ass. So what I think what happened was the cabinet and the President kind of went into shock and didn't know what to do, and Schwarzkopf was waiting for someone to tell him what they wanted done. So when they called I knew exactly what it was and they said, "I need you to come down." I had just landed, so I went out and jumped back in my jet and flew down there.

General Horner knew life would be chaotic once he received that radio call from Washington Center but he did not realize what he would be asked to do. GEN Schwarzkopf requested his presence at MacDill AFB in Tampa, FL to help prepare his brief for President Bush. Gen Horner would soon find out how he was GEN Schwarzkopf's "go-to guy." Trust is important between leaders in the military, and Gen Horner earned GEN Schwarzkopf's trust as no other airman had.

What were the major factors in your relationship with GEN Schwarzkopf that led to his unwavering faith in you? Both to brief President Bush and to serve as the CENTCOM Commander forward?

He didn't like Air Force people, and his experience with Air Force people had been people on the staff, and normally they aren't very quick. He also intimidates people. It started off when he first came to see me. He'd just taken command, and he was touring the component headquarters, and he comes to Shaw [AFB], and we're sitting there just the two of us, and he says, "Well under [Gen George B.]Crist [USMC, CENTCOM Commander, predecessor to GEN Schwarzkopf] you were the Air Component Commander, and I think I'll continue the same thing." I said, "Thank you very much. Now, who's going to be the ground component commander? Marines or Army?" And he goes, "Well I guess I will be." And I said, "Well I guess you'd better get used to me talking to you in a way that you don't appreciate because I owe it to the unified commander to make sure we have the best advice available. So I will take you on as the ground component commander, and I will speak to you in a way that you don't appreciate, and then you will make up your mind, and that's what we'll do." So, I never talked to him in public. I wouldn't even say hello to him in public. He was so insecure about what might result in someone talking to him. So, in a staff meeting, if I wanted to talk to him I'd either whisper or pass him a note in public. In private, I could walk into his office and say, "You're the dumbest SOB I ever met" And he'd say, "Oh why do you say that?" And I'd say, "Because of this, this, and this," and he'd say "You're right."

You played Schwarzkopf for his idiosyncrasies. The other thing I did was we were getting ready to go to a big exercise where country orange invades Saudi Arabia and Kuwait. I wonder where orange is. So, I went down about April, and I gave him a whole briefing about what he should think about in regard to that. One thing, I sat whiplash bingo alert in Vietnam, where the Army in Laos would want air support, but they didn't know when or where so what they'd do, we'd be ground alert close air support (CAS). That's a waste because we'd get right to the end of the period and they'd launch 'em. So, I told him I'm not going to fly any alert CAS. I'll fly every preplanned

CAS sortie, but I will have air over the battlefield, 24 hours, 7 days a week, any one of which can be diverted to support the Army if they need it. It's called Push CAS. So he bought all that. The Navy tried to keep the Navy out of CENTCOM. They had a staff of about 10 guys in Hawaii where the Navy could keep them imprisoned, and the only time I would see them was when they would come to BLUE FLAG exercises because they were CENTCOM Navy assets. But, I knew all these guys and I trusted them and had deep appreciation for them. Schwarzkopf was very good to his component commanders except [LTG John J.] Yeosock [Third US Army Commander] and the reason he was mean to Yeosock was if it was going wrong, he was afraid he'd get blamed so he made sure Yeosock got blamed. Yeosock and I roomed together, and the phone would ring at 0200 in the morning, and it'd be Schwarzkopf, and he'd use the phone even though he could have just opened the window and we'd have heard him.

What was it like briefing the President on such short notice?

It was funny because he had to go to Camp David [Presidential retreat in Maryland] to brief the President. So we worked on the briefing, and it started out that he wanted me to help his staff. His staff didn't want me to help, and I can understand that. So I went up there and sat in the office and thought these guys are a bunch of idiots. So we go in at about 1700 for the practice briefing that he's going to give the President and he gets to the ground thing and it's all Schwarzkopf; he had personally tweaked everything in there. He comes to the air side and they're all acting like idiots, so I'm sitting there and he's starting to blow. I say "Just a minute guys." I'm sitting right next to him, and I say, "Sir if I was going to brief the President here's what I'd tell him, tadat tadat tadat." He looked at me and goes, "Good, you get to brief him." I thought, "Son of a bitch, that's the dumbest thing I've ever done." I had no desire. Anyway, so I go back, and we're going to meet at Andrews [Andrews AFB, Maryland] the next day, the airplane is going to pick me up at Shaw [Shaw AFB, South Carolina]. I'm on crew rest so I just go over to the quarters and I get this phone call, saying Schwarzkopf wants to talk to me. So I go over to his house and knock on his door and he opens the door and he goes, "How'd you get here?" I said, "I never left!"

"You can kill me, but you can't eat me." That's the attitude to have in the Pentagon. What kills people is they get ambitious, they want to get promoted and shit like that. Think about the mission, think about the job.

PLANNING THE GULF WAR

Gen Horner had been the Ninth Air Force Commander but also dual-hatted as the Commander of US Central Command Air Forces (CENTAF) since March 1987. He had spent a significant amount of time in the Middle East in the two years prior to Iraq's invasion of Kuwait. He had a solid understanding of the culture and had built relationships with political and military leaders in the region. GEN Schwarzkopf selected him to serve as the CENTCOM Commander Forward with the primary mission of generating combat power in theater. In a matter of a few days, Gen Horner went from flying what would have been a standard training flight, to briefing the President at Camp David, to establishing a headquarters at Riyadh AB, Saudi Arabia and preparing to take the US to war. Remembering Vietnam, Gen Horner was concerned that the planning would be conducted from Washington D.C.

What reservations did you have when you were told the planning would be conducted from within the Pentagon?

It wasn't ever going to be that way, and Schwarzkopf agreed. He'd been in Vietnam too. So when they sent the team out, [Col John A. Warden came over. So, I was interviewing him to do the job that [Brigadier General] Buster [Glosson, CENTAF director of campaign plans] finally took. We get to the part about bombing the Iraqi Army, and he'd developed this theory that you don't attack Army. Of course, the Army defines war as Army versus Army and everybody else is there to help them. It's the mindset of the Army. So, I understand what he was saying, but the mistake he made was he mirror imaged the enemy, Americans do that. I said, "Sounds good to me. I'll try it but believe me, I'm not going to let the Iraqi Army go unscathed because they may get the idea that they can come fight and I don't really want to have them come fight. I want to kill them where they are." As it turned out, after the war, I had dinner with the head of Iraqi intelligence who escaped in 1997 or 1998, and he was side by side with Saddam the whole war. I asked him, "How come Saddam said he beat us?" He said, "You don't understand, Saddam's only desire was to stay in power. He didn't give a shit about Kuwait. If he could get it fine, if he didn't get it fine. Just as long as he stayed in power, and he's in a country where a lot of people want to kill him. He depended on the Republican Guard to keep him in power. That was his hold over the country." Well, we're taking them apart. He said, "Saddam went into deep depression, he sobbed, sobbing depressed." And then when we stopped the fight he had enough of the Republican Guard to stay in power and he said "I won." Americans never understood that. That's Warden's Achilles heel. In the case of Iraq, if he hadn't mirror imaged Saddam to think that blowing up the legislature is going to win the war or the department of agriculture, which was his theory. In the case of Iraq, bombing the Republican Guard is the way to get them out of Kuwait. Remember, they offered to withdraw from Kuwait before the ground war started. It was just that they wanted to take their time doing it and [GEN] Colin [Powell, Chairman of the Joint Chiefs of Staff] was terrified that the Army wouldn't get into the war. I'm not anti-Army, I'm anti-stupid.

Did you have a chance to influence their planning prior to Col Warden and the Checkmate planners brief to senior leaders Stateside prior to their arrival in theater?

The only thing he brought to the table was the biological, chemical targets that he got from the Central Intelligence Agency (CIA) that had not been previously given to us. They came over a week before that and gave me that briefing. And we'd been planning the attack on Iraq for years. Warden didn't bring much except a flawed concept. Of course, the Army guys hate him because he says, "Don't fight the Army." I said, "Tell me, what should we do about the Iraqi Army?" He says, "If you're as smart as I am you'd know." The room, you could have heard a pin drop. Now, you'd think I'd blow up at something like that. I was thinking, he can't learn and he can't be a team member, that's when he flunked the interview. I require teamship. Now he just showed total inflexibility. Now I have to find a planner because normally I wouldn't have a planner. We'd go to war, and the director of operations (DO) and intel guys would lay out the scenarios. So, that's when I called Buster and said, "Can you get up here?" He was there in the afternoon. I don't know how he did it.

One unique challenge that Gen Horner undertook was the establishment of a coalition. He was tasked with building the coalition of air forces to include identifying and negotiating for basing options, force flow timeline and plan, and tasks and missions for the different nations' air forces. Ultimately, the coalition would grow to over 30 countries and all countries had the same goal of expelling Iraq from Kuwait. However, Gen Horner still had to manage and deconflict differing personalities and operating styles.

As leader of the coalition, what were some unique challenges in building relationships with other countries?

Americans are highly respected, but international partners can get jealous very quickly. So, you never do things that make them jealous. What you do is you do things that make them part of the team. You listen to them; you work very hard at making them part of the leadership team. So, what we did every day, morning and evening, during the changeover brief all the national leaders sat around the table, and it's easy for airmen to do because rank is not an issue. I mean the flight leader is the flight leader, and he may be a second lieutenant, and he may have a colonel on the wing but in flying, the flight lead is the flight lead. In the Army rank is everything. I mean shit, they check date of rank to see who salutes who, very stupid. So we'd sit around the table and everybody talked so the national leader felt that he had legitimacy.

So, the rule of thumb is this, they want us to lead, but they don't want us to act like we're in charge. Schwarzkopf was very good about that. In fact we created, Khalid bin Sultan. Yeosock and I were over there, and I was CENTCOM Forward and Schwarzkopf went back to the States to prepare the forces so we'd talk on the phone at night and we agreed that we needed to get a counterpart to Schwarzkopf. Because in Vietnam the Americans overran everybody, and of course the capital of that country is now called Ho Chi Minh City, so we weren't going to let that happen again. So, we couldn't find an Arab but Khalid was commander of the air defense force in Saudi Arabia, and the Air Defense force goes to the Air Force during times of war, as it should be. And, he was the only royal of their service chiefs.

So, we were in a meeting, the first afternoon I'm there, we're giving the Saudi chiefs, General Mohammed al-Hamad [Saudi Arabia Chairman of Military Forces] was his name, this CIA briefing of what was on the ground up there from the National Reconnaissance Office (NRO). So, first question, he turned to me, "The King has invited you to come to Saudi Arabia?" I said "Yes, he has." And, he says, "How many will you bring?" I said, "Oh probably a couple hundred thousand." And, he says, "Well you aren't bringing women are you?" I replied, "General Hamad, women are integral to everything we do in the military, so even if I didn't want them to come, I couldn't not bring them because it would make our forces ineffective." And the head of the Army, General Joseuf Rashid [Saudi Land Force Commander], was a really good guy and he's sitting there, and he's thinking, "Oh shit." So, he says, "Let me get this straight, you're bringing women over in your armed forces because they are integral parts of the force?" I said, "Yes." He said, "Will they wear a uniform?" I said, "Of course." He said, "Will they drive?" I said, "If their duty requires them to drive, they'll drive." He says, "Will they drive when they're not on duty?" I said, "No, we'll obey the laws of the country we're in." He turns to Hamad and he says, "It's no problem, they're soldiers." And Hamad said, "Well make sure they keep their hair up under their helmet." So, then the next question, I said, "We need to integrate because in Vietnam we didn't do that, we played like we did, but we never did. So we need to make sure that you know everything we're doing and we know everything you're doing and we're all one team." They had just finished this beautiful command center about six stories down underground, under that building we were in. And he, "Ehhhs" and I see he can't make a decision. And this Lieutenant General, Saudi, had a little notebook he's making notes in, he slams that book down on the table and stomps out of the room. I'd never met him before. And so, pretty soon we're talking back and forth, and I say, "Look it, we're going to go down there." You have to be firm sometimes, but I do it nice. So the guy comes back in and sits down, quiet, phone rings. Hamad picks it up and goes to channel B, speaks in Arabic and hangs up. He said, "Let's go down and look at the room." He called his dad who was Secretary of Defense, he's Prince Khalid. I thought, "This is the guy we want on the team." So we had to work very carefully on the Saudi side. Schwarzkopf bought it right away, but the Arabs get very jealous. Finally, what he did, he went to his dad and said, "You gotta appoint me as Schwarzkopf's counterpart, and then I'll retire the day the war ends." That was how he was able to—and he got promoted, and he did retire about a year after the war—and they made him a four star.

Which countries stood out as the best or worst to work with?

The Brits would have been a problem. The British politicians loved to meddle in their military matters. But there was a guy, [Sir] Patty Hine [Air Chief Marshall British Joint Commander Gulf Forces], and what they'd do is the politician would say, "You've gotta tell Horner to do this." And he'd take that, and he'd put it in his desk. It was a good thing.

The Canadians sent over a one-star Admiral. They had two ships and a squadron of F-18s [Hornets]. So, he comes in to see me, and he says, "Well you know, the Canadian Air Force can only fly over the Canadian ships." We cleared that one up right away, the guy went back to his ship and never came out again. And I had a lieutenant colonel from the Canadian Air Force that was great to work with, and we just did whatever needed to be done. The trouble was I needed bomb droppers, and they were only equipped for air defense, air to air. So I said, "If I give you some bombs, your airplanes have all the software and everything to drop them, your pilots are trained to drop them." He says "Ottawa won't allow us to drop bombs, we might kill somebody." So that night, the Deputy Secretary of Defense was coming to dinner. So I got a seat next to her at dinner at the Canadian Embassy. So, we're sitting there, and she says "How are my boys doing?" And I said, "They're doing great, you got the best Air Force in the world. It's better than ours, by far. It's too bad you're going to be the laughing stock after this war." She said, "Why is that?" I said, "Well, I need bomb droppers and your headquarters won't allow them to drop bombs. I'll give them the bombs, I'll give them the loaders. I'll give them everything they need to drop bombs. We have to drop bombs on this enemy." The next day they got approval to drop bombs. So there's a little diplomacy involved. Man, I'm not good at diplomacy. I got some things done that wouldn't have otherwise gotten done.

What was a typical day like in Saudi Arabia? Just to give us a sense of the scope of responsibility you had and the breadth and depth of operations you were in charge of?

You worked your ass off. Except I'd go flying, I'd go visit the troops. I'm a big believer in visiting the troops, in being present. [General Wilbur L.] Creech taught us that. Creech would only let us sit at our desks for three hours a day, max. And boy I tell you, the paperwork piles up but you get up, and you walk out the door. It's amazing what it does when people get to see you and know you're around, and they talk to you eventually. And their bosses get nervous. Unless their bosses are good leaders and if the boss is messing up then he wants someone to know.

LESSONS LEARNED

In 1965, Gen Horner was temporary duty (TDY) at McCoy AFB in Orlando when he received a call to return home because he received orders. He left the next morning for Korat AB, Thailand and what would be his first combat tour. One memory from that tour that was burned into his mind was sitting near his jet and waiting to be called to provide CAS for ground troops. In his mind this was a waste of combat power, once called he would get airborne as quick as possible and support troops fighting beneath the jungle canopy. The situation did not allow for effective CAS because he could not see to the jungle floor and ended up just bombing the jungle. Gen Horner created Push CAS for employment in DESERT STORM as a direct result of ensuring he did not keep pilots on alert when they could be employed.

Looking back on it, how effective was Push CAS? Would you do it again the same or make changes?

Well, we never got a chance to do CAS. There was never a ground war. I told them before the ground war started, I said, "There's no target that's worth your life. So if you can't get in, bad weather or something, we'll go back tomorrow, we'll get it." When the ground war started, I said, "Now, if a guy on the ground needs close air support, I don't care what you need to do, you do it. I don't care if you get killed or not. That has nothing to do with it. If they need it, they need it." Well what the Army learned to do, they'd come forward until they engaged the enemy, they'd back in and then they'd call in air to kill the target. It's not CAS, it's interdiction.

Upon redeploying from Vietnam, Gen Horner volunteered for Wild Weasel training. The Wild Weasels used a new technology to locate surface-to-air missile (SAM) radars and attack using radar homing missiles. After completing Wild Weasel training, he found himself back at Korat in 1967. Weaseling proved to be very challenging at times because he was deliberately trying to fly near SAMs so they would turn on the radars at which the Wild Weasels would be able to engage with radar homing missiles. Conversely, weaseling for B-52 [Stratofortress] employment in Vietnam was drastically different.

People get in their own little corner; our Air Force is as bad as the Marines, Navy and Army, and we're just as stupid sometimes. Particularly the B-52s in Vietnam. You know what I used to do when I'd be weaseling for a 52 strike? Smoke cigarettes and read the stars and stripes, on autopilot. They had one lousy SAM site down at Fingers Lake out in the south part of North Vietnam. And, the 52s would come in from Guam, and they'd drop their bombs in the jungle and do a 180 and go back out over the water. I'd be sitting their weaseling, and my electronic warfare officer (EWO) would be looking for the signal to come up, and if it came up, we'd put a strike down their throat. But, that changed on Linebacker II. The boys got to earn their awards and decorations.

Two tours and over 100 combat missions in Vietnam left a lasting impression with Gen Horner about how wars should be fought. Like many senior leaders serving in the military in 1990, their most formidable years were their time spent fighting in Vietnam. There were numerous mistakes in Vietnam that Gen Horner would not repeat in the Middle East; the most important was the targeting process and approval. He made sure that the commanders in theater, fighting and leading the war, received proper strategic guidance, and understood that the political objectives would direct the military action. Civilian leaders in Washington, D.C. were not going to be involved in target selection the way they had in Vietnam. The President, Secretary of Defense, and Chairman of the Joint Chiefs of Staff had developed an environment where this was not only possible, but it was the way the US was going to fight this war.

What lessons learned from Vietnam did you apply during DESERT STORM?

Integrity. The first thing we lost in Vietnam was our integrity. I'd come back from a mission; they'd say, "What'd you hit?" "Oh 100 percent on the target." "What was your release altitude?" "4,500 feet." "What was your dive angle?" "45 degrees." "What was your airspeed?" "450 knots." That's all I ever told them. Of course, we'd get up there, and there wouldn't be anybody up there but us so hmmm...and once you lose your integrity in your military you've lost everything.

The other thing is we're in it until it's over. Vietnam in the south they'd have their calendar crossing off the days and in the north the missions. I didn't think that was right. I think you ought to be committed. In Vietnam, I don't know the relationship between [Robert] McNamara [Secretary of Defense] and the President [Lyndon B. Johnson], Joint Chiefs of Staff and [GEN William] Westmoreland [Chief of Staff of the Army], but obviously, it was deeply flawed. Goldwater-Nichols [Department of Defense Reorganization Act of 4 October 1986] was the best thing that ever happened. The best thing that ever happened to me was George Bush and Dick Cheney because Cheney always knew everything that was going on. He'd question in large detail, but his questions were not designed to show that he was smarter than I am and he wanted to know the answer. A lot of times we'd say things, and I'd say, "I have no idea" and he'd say, "If you don't know the answer, don't give an answer." President Bush, the only two things at Camp David he talked about were loss of life and the coalition. I think that's because he was in WWII. He meant all lives, both Iraqi and ours. So, we planned everything trying to minimize loss of life. And I thought we did a pretty good job. Colin came out and said we killed 100,000 Iraqis and I said, "My guys didn't see 100,000 dead Iraqis." So I was in St. Louis and I got asked how many did I think by an anti-war person and I said "Well, they started off with 500,000 of which they only employed 400,000, half of which deserted, 200,000, for everybody that deserted, some came towards us, 88,000-100,000 down there. So that leaves 100,000 unaccounted for so if somebody told me we killed 10-15-20,000 I'd believe them, but we didn't kill 100,000." My aide came in and said, "Powell just said we killed 100,000." I called Pete Williams who was spokesman for Cheney and I said "Pete, I may have stepped on it big time. I just said 20,000 max." He goes in and comes back out and says, "Boss says, you were there." That's the kind of guy Cheney is. When I see him to this day, he's just down to earth and just as nice as he can be.

The US led coalition of 34 different countries conducted 38 days of air operations leading up to the 24 February 1991 ground invasion. It took only 100 hours to expel Iraqi forces and restore the sovereignty of Kuwait. Gen Horner led the air campaign that flew over 100,000 sorties that dropped over 88,000 tons of ordnance.

Were you satisfied with how the Air Operations were executed?

Well, we screwed up a bunch of things. Never satisfied. But we did one thing, Vietnam again, we made sure that the individuals executing the air tasking order (ATO) had a say. They didn't like something they could call up and say, "I don't like it." And I'd say, "Fine, what do you want to do?" Because I could have done Vietnam better than the guys who planned it in Saigon. Because I'd been there. I knew what to do and what not to do. You're a captain. If you know more than the generals you get in trouble. My guys came to me and said "How come you knew we weren't supposed to go low-level?" I said, "On 22 July 1965, we went against the first SAM raid ever. We went in low-level and we had six guys shot down, three of them killed and one recovered as a prisoner of war. I was on that mission." I said, "Man, we aren't ever going to do this shit again. Came back and had holes in the top of the airplane." So, I was uniquely experienced for that job. When I was a two-star on staff at Tactical Air Command (TAC), they always had to send a general to be the air component commander at these big joint exercises; they'd always send me because I was the planner and I didn't have anything else to do. So all my guys knew everything, I didn't know everything. I worked with Gen Al Gray [Commandant of the Marine Corps], a Marine and a really good guy, the XVIII Airborne Corps Commander before LTG Gary Luck. Gary Luck who was also a great American. These are great guys. They have a little trouble with the infantry and armor guys. They're a little slow but the Korean and airborne guys, see they have to have air or they die so they are a lot more easy to work with. The armor guy is always thinking, "Here's how I'll use my air." There's two armies, there's Europe Army and there's Korea Army.

Targeting, I would not allow any targets to be nominated. People could send me information on targets all they wanted, but every target was selected in the black hole and approved. And that isn't that hard. What you do, for example, a lot of the air, we would destroy artillery and armories. That was Schwarzkopf's desire, and it was the right thing. Kill their killing mechanisms, not their troops because their troops weren't worth a shit anyway. Except the Republican Guard. So when you're allocating air towards deployed Army forces you can allocate batches. So Schwarzkopf at night, since he couldn't get any good intel when he did, he had on the wall, up along the top he had every Iraqi Military Division. Then he'd keep track of, he'd get boss stuff, morning reports, intercepts stuff like that so we'd have some idea up there. So, he'd keep track of how they were doing. So he'd say, "Let's put so much on that one. Let's make these units the prime and such." And so we'd put the weight of effort on those. You don't get into individual—what you do, you tell the dogs of war, "Boys get your ass up there and if you see a tank kill it in this area." And then we got the killer scouts because the Army would depend on week old intel. So, the first week I bit the hook and sent them to where the Army wanted them to go, and they get there, and there wasn't anything there. So what we did, we sent an F-16 in, and he'd have a kill box, we did that a little bit in Vietnam. We called them fast forward air controllers (FACs), but they aren't FACs so we called them killer scouts and they could kill too because they had guns and bombs and such. See you're always seeking efficiency in warfare because war is very inefficient. The other thing an airman always does that the Army doesn't necessarily do is—the airman's job unless your homeland is being under attack—the airman's job is to manage attrition. If you aren't losing any airplanes, you aren't doing shit, and if you're losing too many you aren't going to be able to finish out the game. So, you have targeting, but one of the factors on targeting is attrition. So, you have to protect the force somewhat. But in tanks it's the same way, you have to watch the attrition because it's vehicle oriented like air.

The general consensus is that DESERT STORM was one of, if not the most successful military operations in the history of our country. Do you agree?

I have trouble with the words. What's your measure of success? Time? That was good. Low casualty rates? That was good. The reason DESERT STORM got ballyhooed so much was the American people had a dirty taste in their mouth from the way they treated the military in Vietnam. The reason they had a dirty taste in their mouth from Vietnam was we were so stupid in the way we fought it. I mean we sacrificed people for no gain. Americans don't like that. So what happened is when we finally did something right, then they were euphoric. I think that DESERT STORM is inflated; the good feelings of DESERT STORM are inflated. But, in a way, it wiped out a hangover they had over the way they acted in Vietnam. So, nobody knows what's going on. Particularly in Saddam's country where you're not allowed to think. I discovered that in '89, I was in Pakistan, and the Pakistan military gets their retirement funds from investments. They have a beautiful hotel in Karachi. I was having dinner at that hotel with a fighter pilot from Pakistan who had been assigned to Baghdad teaching the Iraqi Air Force, and I said, "What are you doing back here?" He said, "I got thrown out of the country." I said "Why?" He said, "Because I started teaching the Iraqi's tactics and the Russian's fired me." Because they teach them takeoff, land, and follow Ground Control Intercept (GCI) control. And, man, that gem cost them dearly. What'd we take out first? Command and control, then they were screwed.

What could we have done better?

We have to train our intelligence people. The way we train our intelligence people, we have these little games, and they write a script, and they get up and brief everybody and everybody says thank you very much. "Now here's what we're going to do, you bring your tanks up over here, and I'll bring my machine guns here, and I'll put my airplanes here."

So, the intel guy, all he's doing is first just writing his own reading like he doesn't have screw all to do. The second thing is if they make a mistake we kill them. So they don't ever make mistakes and they say, "The indications are that it's probable that..."

nothing concrete. And then you go to war, and you suddenly realize that you don't know what to do other than what intel tells you. They are fundamental—they're the basic elements. I had to train intel people. I had a Navy commander who was in charge of the transportation system. So we're knocking out bridges, wanted to isolate the battlefield. So he'd get up and say, "Yesterday we'd have..." I'd say, "I don't give a shit what we had yesterday. What does that have to do with anything? What do I want to do tomorrow?" So eventually, I got him so he'd say "Well sir, tomorrow you have to knock out this bridge and this bridge. It'll cut throughput 32 percent." And I thought, "You lying sack of shit, but I love the way you think." And that's what you have to have. You have to have intel people who have balls, and it's hard to find them because they smother anybody who has any kind. I had to fire the Rivet Joint [RC-135V/W reconnaissance aircraft] guy. We're over there, it's hotter than blazes, we're eating paper cups, and everything is terrible, and he comes in and says, "Well, I'm going to take my people and go home." This is like in September. I said, "Why?" He said, "We're tired." I said, "We're all tired." He says, "Well I don't work for you." He's a Title 50 not a Title 10. He works for National Security Agency (NSA). I said, "You're right, get your ass out of the theater immediately." So, now I have to find a commander, and we get this guy, and we call him Moody Sooter. His name was John Lewis. We called him Moody Sooter because he's a wild thinker and that's exactly what I wanted. So we can collect the right kind of information and make something out of it and send it back. We're at a staff meeting at the table, and he said "Sir, we'd like to introduce our new commander for the rivet joint, it's John Lewis. Stand up John." And I said, "You're almost as fat as I am." And he looks at me and says, "Almost." Everybody in the room went silent, and then everybody laughed. So intel, intel needs work. They worry about the bomb damage assets, that's the ops-antis. They're always trying to measure everything. Now what you want to measure in war, it's not activity, you want to measure output. "I don't care how many bombs you drop, that doesn't tell anything. But how many artillery pieces do they have left and what are they doing because they can't group them up. What impact does that have?" Stuff like that. We're not good at, Creech taught us to measure output, not activity.

Besides the organization and control over the Combat Search and Rescue (CSAR) efforts during DESERT STORM, what went into the decisions whether to or not to launch a recovery mission?

It's a service responsibility to do the rescue of that services' members. So if a sailor falls off the destroyer, it's up to the Navy to put a boat in the water and get him. What they do is, they send that up to a central headquarters under the unified commander. I operated that headquarters, because I was the only one who had communications throughout the theater. So, a downed person, a missing person, it was beyond the capability of that particular unit to rescue him, I'd get it, and I'd decide whether we could rescue him or not. But, I did that in the name of the unified commander not as the air component commander.

We had, Maj Rhonda Cornum, she got shot down trying to do a rescue that was valiant but probably ill-advised. I would never criticize. You do what you do at the time. I never want to inhibit people from taking matters into their own hands on the battlefield. But I wouldn't have ordered her to go up there and get it.

Most were airman who were shot down over the battlefield. We didn't have a lot of capability to go in there, and I know the war wasn't going to last that long. So, I knew we weren't facing a Vietnam situation so therefore I would err on the side of caution rather than—and quite frankly we didn't have a lot shot down. That helps a lot.

LEADERSHIP

Gen Horner served over 30 years in the Air Force, fought in multiple wars, and commanded thousands of Airmen. He had the privilege of serving under great leaders and having mentors during his career. One of these mentors for Gen Horner was Gen Wilbur L. "Bill" Creech. Gen Horner first met Gen Creech, the former commander of TAC, at Luke AFB, Arizona in 1979.

Did you have any enduring mentors you sought advice from during your career? What qualities did you try to emulate?

Creech was a mentor, and he took that job very seriously. He felt that part of his job was always to train his replacement and he imbued that to us as the wing commander. As a result, whenever I had a really sterling person I'd start looking for a job for them someplace else. Because word gets around and it's hard to hire good people if they think they're getting stuck there making me look good. Creech taught me that. Creech was awesome. I hated being around him; he was a pain in the ass. In fact, there's a book about him, about what I would do when we went to commander's conferences. What I'd do, like a force field, I'd keep track of where he was in the room and stay away from him. Because I wanted to get drunk and have fun. He read that, and he laughed his ass off.

How much did the training prior to 1990 account for the dominance of the Air Force in DESERT STORM?

You start off as a second lieutenant and you learn how to fight. And first lieutenant and captain you learn how to lead. Major and lieutenant colonel you learn staff. As a colonel, you learn joint. Sometimes career people don't like to go outside their own service because they're afraid they'll lose their place in line for general or something like that. You're a careerist if you start thinking like that. I mean, I know you're screwed because you came to Maxwell [Maxwell AFB, Alabama] instead of Leavenworth [Fort Leavenworth, Kansas] [talking to Tommy]. My son went to Leavenworth. He was an A-10 [Thunderbolt II] pilot, they thought he was God. I never went to school at Maxwell [AFB] and it pisses them off big time because sometimes they ask me up there to talk and I'm not their idea of what an Air Force officer ought to be.

When I was on the book signing tour with Tom Clancy, we went out to Colorado Springs [Colorado] because it's kind of a military town. So Robin Olds [Brigadier General, fighter pilot] lived out there, and we were going out to dinner, the Air Force Academy wanted me to come to dinner. So, I said, "Is Robin coming?" And he said, "Oh no, we can't have him come." I said, "Why?" He said, "Well, he'd get drunk and start swearing." I said, "If he doesn't come then I don't come." And Clancy wanted to meet Olds. So, he sits down at the opposite end of a big ole table across from Olds. Clancy liked to try to piss you off the first time you meet him. So, he goes to Olds, "Are you really as good as everybody says?" And Olds says, "No, I'm better." That broke the ice.

What is one word you would use to describe the most important aspect of leadership?

[GEN George C.] Marshall [Chief of Staff of the Army] was talking to [GEN Dwight D.] Eisenhower [President of the United States and Supreme Commander of the Allied Expeditionary Force in Europe] after the invasion of France and they were going over there, and they were staying there. And Marshall turned to Eisenhower and said, "What's the most important thing you look for in a leader?" and he said, "Courage." And that really has to do with having the courage to make a decision. And what is a decision? A decision is acting when you don't know the answer. Most of the time you know an answer, or you have clues so that's not a decision. It's not my idea it's Marshall's idea.

I think, selflessness is important and I don't know how to describe it. You wouldn't describe GEN George Patton [Third Army Commander] as selfless, but in a way he was, he was willing to die. So, they say concern for the troops. Well, the trouble with concern for the troops is, in Vietnam, we tried not to kill them, and we killed a lot more than we should of, of our people. Sometimes you have to send boys into hell.

We went down to Diego Garcia [AFB on Diego Garcia Island in Chagos Islands in the Indian Ocean], and we were briefing them on what they were going to do when the war started. And I said, "You might look to your right and look to your left because that person isn't going to be there." And these are B-52 people; they had led a very strict and sheltered life other than Linebacker II. And the guys were shocked. But it turned out we lost one airplane and it was an accident rather than a shoot down. They came to me one time. They said, "The 52s are not going to hit this target." I said, "Why?" They said, "There's SA-6s [GAINFUL, SAM] there." And these guys are at 40,000 feet, and a six doesn't go 40,000 feet. Plus, they have jamming and stuff like that. So I said, "Would you please tell the wing commander that the target is north of where he is and to get there as soon as possible?" The opening shot of the war was the cruise missiles launched off the B-52s out of Barksdale AFB, Louisiana. And that's a public relations (PR) plow but none the less, I'm all for letting somebody make some money. Like, we let the Apaches (AH-64, attack helicopter) shoot the radars up like that was a big deal. And Schwarzkopf went livid when he found out it was Special Operations helicopters, so that's why we put the Apaches on their wing. That solved it because the Army got to fire the opening shot of the war and so Schwarzkopf was happy. But anyways, the 52s coming out of Barksdale, they hit Egyptian airspace, and the Egyptian controller calls them and he says you are not cleared. And, the B-52 lead was going to turn around and go back to Barksdale. Fortunately, the wingman said disregard, go channel such and such.

We often learn more from failure than from success. Do you have any specific examples of negative characteristics in a leader that you made sure you didn't exhibit?

I worked for a wing commander who was vicious, mean spirited and he was terrible. So we had a big wing, we had F-104s [Starfighter], F-4s [Phantom II], F-15, F-5s [Tiger], big wing. And he decided to divide it into two wings. I was his vice commander. He was so disruptive in nightly meetings I wouldn't sit at the table. I would sit behind him in the chair because I wasn't going to be part of that. But I kept my mouth shut and just sat there. The day before they stood up the new wing, they had given me the job as the vice commander to divide up the assets. It was all one maintenance. Well, I had F-4s and F-15s so I had to divide up. The F-4 was the older airplane so I put the best guys to take care of the old airplane and stuff like that. The day before they stood up the new wing, they fired him and made me the F-15 wing commander. And I'd been flying the F-4s, he wouldn't let me fly the F-15. That is a stroke of genius. They led me to believe I was going to be the F-4 commander and had me organize everybody who'd go where. So anyways, I ran into whenever I didn't know what to do, so I'd think, "What would he do?" And I'd do just the opposite and it was always right. So negative role models are wonderful.

What can our generation of officers learn from the wars post 9/11 in Iraq and Afghanistan?

You're never going to be perfect so you might as well relax. Are you having fun, or not having fun? If you're not having fun then go find a job someplace else. I loved being in the military. I loved flying airplanes. And I loved being with the people I was with. So I had no desire to go with United Airlines. I could have made a lot more money, but I would have been miserable. If you're not having fun, get out of the military. If you're having fun, stay. It really doesn't matter what rank you make. A person's worth is not based on rank. Don't get confused about that because some of the dumbest shits I know are four-star generals. Have fun. Work hard. And you're going to have bad days. I think the worst feeling you have though is when you screw something up, and you knew you screwed it up. And if you didn't know you screwed it up then you're probably dumb enough.

Another thing I would tell wing commanders, "You want to get promoted to General?" "Yeah." "Then just fire the right NCO [Non-Commissioned Officer]." I said, "Look, NCOs make the Air Force run. You got some of them that are scraping by, you got some of them that are doing all the hard stuff, and they all know who's who. But, they aren't ever going to tell you because you're the officer. So start firing NCOs, and if you fire the right ones, the ones that are skating by, they may be the best dressed. They may say 'yes sir, no sir.' They may salute the sharpest. But, if they are the right ones to fire, the workers will kill for you. If you fire the wrong guy, he may have hydraulic fluid all over him but that's because he's teaching them how to change hydraulic pumps. You fire him and the rest of them will sit down and wait for a new guy to come in."

Another thing about the Air Force is we stick 2nd Lieutenants out on the flight line and there'll be a Chief Master Sergeant out there that's been on the flight line it for 30 years. And you'll wonder, what am I doing here? Well, I'll tell you what you're doing there, you're there to learn. But the role of the NCO is enforce discipline in the enlisted force and there'll be time when you have to depart from the book. Now he can do it, and is willing to do it, but if you do it, then he becomes compromised as an enforcer of discipline. So we let that second lieutenant who doesn't know shit about discipline, hopefully with the advice of the NCO, deviate from the book and if it gets screwed up, then we just fire him. He's a Lt, they're expendable. But don't ever put an NCO in the position where he doesn't enforce the rules, because that's his job.

Gen Horner added some advice for the ACSC Class of 2017.

Go have fun at school; it's a pain dealing with a lot of dweebs running the place. I went to Command and Staff College in Norfolk, and my running dog buddy was a Marine who had been at Inchon [Korea] as a private. We called him "Stainless Steel." He'd pass out at attention and be laying on the floor like that. We had a colonel who was teaching of course, and he didn't know what he was talking about so both of us, for different reasons, attacked him. He left the room crying. He was Air Force. His boss was an Army colonel, and he called us in and chewed us out, and he was going to throw us out of the school. The guy was wrong.

What has been the most rewarding leadership experience you had in your career?

Well, the highlight of my career was DESERT STORM. You know during DESERT STORM I wrote my letter of resignation. It's true. I wish I had it, but I lost it. I was not going to let Washington dictate buffer zones and things like that. But it never came to that. In fact, we were at a stamp issue, a DESERT STORM memorial stamp, and they had us up on the stage with Cheney. And I said, "Did you hear about us shooting down the two Iraqis?" And he said, "Yeah I did, I wish you hadn't done that." I said, "So do I, the guys missed the street sign. Did anybody offer any suggestions on how to keep that from happening?" He said, "As a matter of fact, I recall I got a staff summary sheet on suggestions that would minimize the chance of that happening. And I remember, I wrote on the bottom, they'll know what to do in theater and sent it back." It was after the war that I learned that. See, I was waiting for Schwarzkopf, I'd go to the meeting that night. Integrity, when we shot them down in Vietnam, we'd just replot the shoot down. I called Schwarzkopf and said, "Good news, bad news. Good news, we shot down two Iraqi's, bad news is we shot them down in Iran." And I know he'd tell Powell and Cheney. So I sat there that afternoon, and I thought, you know it'd be the Assistant Secretary of Defense (SECDEF) who'd say "We can't have that." Figured something like that. So, I wrote down all the arguments I'd have.

Then I realized that if it comes from the SECDEF, Schwarzkopf isn't going to fight it. I mean, what the hell? And that's when I wrote the letter of resignation; I was going to give it to him. That was all he could do. Went to the meeting that night, nothing happened. Next night, nothing happened. Next night, nothing happened. Finally, I lost the letter. That's why I have such great respect for Bush and Cheney. Powell is a friend of mine; we went to War College together. He's a politician. He's a great politician, probably the best politician that ever lived in Washington, D.C. But he's not a warfighter. He was terrified they were going to give me CENTCOM after the war. Cheney wanted to give me CENTCOM. And I went to Powell, and I said, "I don't want CENTCOM, I've had enough of the desert."

Gen "Chuck" Horner is an Airman's Airman. He is a command pilot with over 5,300 flight hours in the F-100, F-105, F-4, F-15, and F-16. He commanded at all levels of the Air Force and served in both the Vietnam War and Gulf War. Few military leaders have given of themselves to their service and country like Gen Horner has. He continues to impart his knowledge, experience, and lessons learned on young leaders in the US military. He has a fighter pilot aura about him. He loves to tell stories and willingly addresses his shortcomings and mistakes in an attempt to make our military better. No matter what service or position one serves, Gen Horner provides valuable lessons that stand the test of time and can make any Soldier, Airman, Sailor, or Marine a more enriched leader.

- Interviewed by MAJ Thomas Bramanti, US Army, and Maj Vanessa Wilcox, US Air Force

Lt Gen Charles Horner, Commanding General, US Central Air Force, receives a plaque from Col Ahmed Al-Kuwari, Qatari Base Commander, during his visit in the aftermath of Operation DESERT STORM.

Lt Gen Horner, 9th Air Force Commander, conducts a prefight cockpit check in an F-15E Strike Eagle.

"The bottom line is not dollars per aircraft but overall capability per dollar."

– Gen Charles Horner.

In Southwest Asia, Gen Horner directed all US and allied air operations for Operations DESERT SHIELD and DESERT STORM, 1990 to 1991.

Civilian and military officials pose for a group photograph prior to discussing US military intervention in the Persian Gulf during Operation DESERT SHIELD. Dignitaries (L-R). Front Row: P. D. Wolfowitz, Under SECDEF for Policy; Gen Powell, CJCS; R. Cheney, SECDEF; GEN Schwarzkopf, CC/USCENTCOM; Lt Gen Waller, Deputy Chief of Staff, USCENTCOM; and Maj Gen Johnston. Back Row: Lt Gen Horner, CC, 9th AF, TAC; Lt Gen Yeosock, CC, 3rd Army; Vice-Adm Arthur, CC, Seventh Fleet and Col Johnson.

37TH TACTICAL FIGHTER WING

Gregory A. Feest

F-117A PILOT | DROPPED FIRST BOMBS IN DESERT STORM

Major General (ret) Gregory A. "Beast" Feest has the unique distinction of having dropped the first bombs in two different operations: Operation JUST CAUSE in Panama and Operation DESERT STORM. He began his military career as a navigator in 1978 after receiving his commission from the University of Wisconsin, Madison Reserve Officer Training Corps program. He later became a pilot and held a variety of flying assignments, including commander of operations and maintenance at one of only two operational F-117A Nighthawk squadrons. He also commanded the 479th Flying Training Group at Moody AFB, Georgia, the 379th Air Expeditionary Wing in Southwest Asia, and 19th Air Force at Randolph AFB, Texas. He retired in 2012 as a major general after serving as the Air Force Chief of Safety. Feest is a command pilot with more than 5,600 hours in myriad aircraft, including more than 800 combat hours earned during Operations JUST CAUSE, DESERT STORM, IRAQI FREEDOM, and ENDURING FREEDOM. While a major assigned to the 37th Tactical Fighter Wing, he led the first ATO that initiated the attacks of Operation DESERT STORM. His target was the Radar Intercept Operations Center (IOC) at Nukhayb, Iraq, that linked Iraqi border radars to the Air Defense Headquarters located in Baghdad, Iraq. Although DESERT STORM was not the first operational use of the Nighthawk, this mission was distinctly different from Operation JUST CAUSE. Panama had virtually no air defense, but Iraq was considered one of the most heavily defended countries in history. On 17 January 1991, Feest released a single Guided Bomb Unit (GBU)-27 on his target, and the 2,000 pound, laser-guided bomb penetrated deep into the Iraqi air defense network's IOC. The strike was part of a well-synchronized mission that quickly followed the first attacks from the US Navy's Tomahawk cruise missiles and Task Force Normandy's helicopters. Close on the heels of these attacks, coalition air forces began mercilessly pounding countless Iraqi targets. Through the remainder of the Iraqi campaign, the F-117 would continue attacking the most difficult targets that required the highest degree of precision. Accounting for only 2.5 percent of fighter assets, the Nighthawk was credited with destroying a remarkable 40 percent of targets in the first three days of the war. Feest completed 39 combat missions against varying critical targets throughout the war. Surprisingly, despite flying some of the most dangerous missions of the war, the F-117s did not record a single incident of battle damage.

While growing up in Racine, Wisconsin, Feest often watched Navy fighter jets fly training missions above Lake Michigan. He knew at an early age he wanted to be a pilot. However, the Air Force did not need many new pilots after the Vietnam War ended, and the Air Force offered him a navigator slot instead upon graduation from college. After spending two years as a navigator in the F-111 Aardvark, the Air Force accepted his application for pilot training. His first assignment as a pilot was in the F-15C Eagle at Langley AFB, Virginia. He did not know at the time that his experiences and accomplishments in the F-15 would prepare him for his next assignment in the secretive and selective F-117A program where he would eventually cement his place in history.

OPENING NIGHTS OF DESERT STORM

How were you selected to lead the first night's strike package in the F-117?

I got asked to join the F-117 program because I did pretty well in the F-15. I flew the F-15 for less than three years, and I became a Standards and Evaluations Flight Examiner (SEFE) and also a flight commander as a captain. Most of them were majors at the time, and I had won some Top Gun awards there. When I got into the F-117, for me it was always super easy. It was like playing a video game, I always said, only it was free. I thought it was always pretty easy and I did very well. I was the wing Top Gun for at least two years. I got picked to be the mission commander and lead for the JUST CAUSE mission. That mission actually was a success even though the Army leaked that the F-117s were involved. Nobody was supposed to know. The Army told us we couldn't bomb the barracks, so we didn't, yet the bombs went where we aimed them. My director of operations then picked me to be the lead for DESERT STORM. I led all the F-117s out of Tonopah, Nevada, to get them over there. Then my squadron commander and wing leadership picked me to be the first guy. I think part of it was because they wanted to prove to the world that I didn't miss in JUST CAUSE, which I did not do. They wanted to give me another opportunity to excel.

Did you have a mission planning cell that prepared the first night's attack for you, or were you involved in this mission planning?

When we got there, it was August, and DESERT STORM didn't start until January, so we had many, many months in the mission planning cell to prep for battle. When we first got there, we didn't know when the missions were going to start, so a couple guys were in the planning cell. I was not because I was going to be flying the first wave. The pilots that were selected for that first go were not part of that because they had to be in crew rest all the time. When we figured out the battle wasn't going to happen [immediately], I was added to the mission planning cell. There were four of us, and later there were six of us. We would spend time going up to Riyadh, Saudi Arabia, working with the black hole [planning room] folks up there, [Lieutenant Colonel] Dave Deptula and Brigadier General Buster Glosson, planning our role and what we were going to do in the F-117.

The F-117A stealth technology made the aircraft low-observable, but not invisible, to enemy radar. Nighthawk pilots, known as Bandits, would reduce the aircraft's radar cross section as much as possible prior to combat by retracting the jet's antennas and removing its beacons, a process pilots referred to as "stealth up" the airplane.

How do you mentally prepare, knowing you are going to go out there on night one in a country that is heavily defended with SAMs?

Not many Air Force pilots had any combat time back then. The Vietnam veterans had all retired, so there weren't any fighter pilots that had combat time. I don't call JUST CAUSE combat. In fact during JUST CAUSE we didn't even "stealth up" the airplane. In any case, we didn't have combat time, but we all had trained at Nellis AFB, Nevada, and at other exercises including RED FLAG. Any pilot in the Air Force says we're going to train the way we fight, and that's what we did. We practiced, we practiced, we practiced, and when it was time to go, we would go. To us, it was the same thing. It wasn't that big of a deal because that's the way we trained, and that's the way they still train today.

As a consequence of putting the aircraft in stealth mode, the aircraft entered a communications-out status, preventing pilots from sending or receiving radio transmissions until the pilots extended the antennas again en route to home. Additionally, the aircraft at the time did not have a GPS to guide pilots to their destination or a radar warning receiver (RWR) to warn the pilots if they were being tracked or fired upon by SAMs. To reduce their risk, the pilots flew along what they called the black line—a flight path that mis-

sion planners believed would keep the pilots away from the most dangerous air defenses.

Although you train like you fight, nothing ever goes as planned once you actually do launch. Can you detail your experience that first night, once you actually did go out there and execute this mission?

During DESERT SHIELD we would fly missions up to the Iraqi border. We did everything comm-out. The only time we talked to anybody was post-mission or we talked to the boom operators. It was all timing and visual, even though it was at night. Then we would fly up to the Iraqi borders and do some orbits along the border there and then we would fly back home. All of this was to desensitize the Iraqis to see that we would be flying up there. We'd do this once a week or once every ten days or so. When it came time for the actual missions when we started the war, we did the same thing, only this time before we left the tankers we stealthed up. Hopefully, they didn't know we were doing it. If the early warning radars couldn't see us or pick us up, then they couldn't hand us off to their target tracking radars, which could shoot us. We didn't know as pilots whether this was going to work, but we trusted the engineers. We knew what our radar cross sections were supposed to be. We mission planned to avoid the threats that might be able to see us better than other radars. We flew what we called the black line; however, that black line always had some red time; meaning, during those times of flight we could be seen by radar and possibly shot at.

My mission was to take out an IOC so that radar would shut down and the F-15E [Strike Eagles], F-111 [Aardvarks], F-16 [Fighting Falcons], and everything else could come through, and radars wouldn't be targeting them. I had two GBU-27s; the first was going to go on this target. I hit it and saw the doors blow out and all the smoke come out through the different ends and through the vents, so I knew I hit it. That was a good thing, and it felt pretty good. I had to do about a 270-degree turn, so as I turned, I looked down. I had a wingman about one minute behind me, and I wanted to see if I could see his bomb hit. I figured I had nothing better to do. When I looked down there, I saw fireworks, sort of. All these red lines flying through the sky, and a lot of it was coming at me. I go, "What did I hit?" Then I figured out, "Oh, that's anti-aircraft artillery (AAA). Those are tracers." I had never seen them. I had been to Nellis, did RED FLAG, and talked about it. You always talk about triple A, but you never see it. It's never really fired at you in training. When I figured out what it was, I didn't care about watching my wingman's bomb anymore. I just looked out where I was going and put the throttles up to full mil-power and climbed out of there as fast as I could.

We each had one more bomb, and we were going out to the west where the SCUDs were set up to launch at Israel. We were worried about that, not just mission planners and pilots, but also our senior political leaders. As I went out toward the west, I looked toward Baghdad, and the sky was lit up. When my bomb hit, they figured out something was going on, and they just lit up the sky with everything they had. I could see SAMs launched; I could see triple A. That's what we always described looked like the fireworks show at Disneyland. The sky was just lit up. It was all barrage fire. They didn't know where anybody was. That was off to my right wing, and I thought, "Boy I'm glad I'm not going there." I looked out front where I was going and it looked the same. It wasn't as big an area where I was going compared to Baghdad; however, I had to go right into the heart of it because that's where my target was. Our techniques for avoiding triple A and SAMs was to lower the seat and don't look outside anymore because the best radar cross section we could achieve was straight and level and doing what we had planned. If you banked the airplane or opened the weapon bay door, your radar cross section changed and that's when they could shoot you. I had to go hit that second target, which was a sector operations center. It was a command center that would help target our fighters. I hit that target, and as soon as I did, again I was below 10,000 feet [due to GBU-27 software requirements]. I climbed as fast as I could out of the heart of the triple A and turned south toward Saudi Arabia. Again, my wingman was a minute behind me, and I was just hoping he was following me through this whole thing, but I didn't know because we couldn't talk on the radios.

Was there anything that truly surprised you that you encountered along the way?

As we came out and got closer to the Saudi Arabia border, we would de-stealth. We put out our antennas. I made a radio call to my wingman to see if he was there. We had a set time to communicate before we rejoined and rendezvoused for the tanker. We needed a tanker to get back to our base in Saudi Arabia. I had a kneeboard with all the pilots that were in that first wave. My wingman checked in, and I slowly checked off all the call signs of the other 10 pilots. After a few minutes, I had 12 names with checks next to them, so I knew everybody was on their way home. As we got to our tanker, our antennas out, now we could hear a lot of people talking about fighters that had gotten hit, diverting and things like that. But until we got back to Khamis Mushait, Saudi Arabia, we didn't know if any F-117s had sustained any damage. We got home, and there were other fighters at our base that had gotten hit and couldn't make it to where they were supposed to go. When all the F-117s landed, not one had received any damage, which was a shock.

We figured out that stealth technology worked and that they were having difficulty finding and tracking us to shoot at us. After a week when no F-117s got hit with anything, we knew it was just going to be the lucky shot that would hit somebody.

You mentioned you would lower the seat and not look outside when Iraq would light up the sky with triple A. Is that true, or is that just an expression?

I never actually lowered the seat, but sometimes I would stop looking outside. [laughs] Throughout all the missions, especially low level below 10,000 feet, the triple A was just unbelievable. At night, it all looks like it's coming right at you, even though it may not be. Sometimes you could smell it. Sometimes you could feel the aircraft buffeting. You don't know if it was from the triple A or from turbulence, but everything felt like it was going to hit you. The best thing to do was not look at it. If there was bad weather, everything would light up the clouds and you could see it going over your canopy or going around you. At night, everything looks closer and it looks like it's going to hit you. So that's why we would say we would just lower the seat and stop looking outside and concentrate on finding our targets, tracking our targets, and dropping on them.

Was it pretty easy to find the targets once you got into the target area?

The way we would plan is we would find offsets. Some targets were harder to find than others obviously. We would have photos. Without the photos, we couldn't find the targets. Thankfully we had all that time during DESERT SHIELD to get a lot of the photos of our targets, at least the first few days of targets. We even used spot photography from the French as one of the ways we got photos back then. You'd have to have a photo, then you'd pick an offset like a bridge or a tall building or a baseball field, and you'd use these as offsets to find real small targets. Then you'd just work your way in until you found it and track that.

Is there anything else within that first week or two you got to take part in that you're truly proud of and like to tell the story?

I was picked to go do this, but a lot of other people could have done the same thing. I was just lucky to get picked. My wingman was one minute behind me. He was number two, but he told me once, "No one will ever remember number two." His name was Dave "Dog Man" Francis. I believe he was a captain at the time. There were a lot of unique missions and a lot of missions that were key to the success of DESERT STORM. General [Chuck] Horner decreed after the first few nights when a lot of planes were getting hit down low that the only aircraft that were going to go to downtown Baghdad would be the F-117s. That was because of the threats—the SAMs and triple A—and how intense it was. That was all well and good, except again we had to be below 10,000 feet. We were just waiting for the mission where one of us would suck up a lucky bullet and take out an airplane, but it never happened. I think we flew 1,271 sorties without any damage, which is something. There were different missions, but they were all exciting. I'm glad I didn't fly night two. I mission planned. Night one was scary as hell. In between my first target and my second target, I don't know where my videotape is, but into my videotape, I said goodbye to my wife. She was at Khamis [Khamis Mushait, Saudi Arabia] on the ground because she deployed, as well.

After that first flight, you learned the President and the Secretary of Defense thought that you all could lose up to half of your aircraft on that first night's strikes. Did this change your perspective going forward into your mission planning or did you just continue mission planning as you were?

The engineers, the men and women who made the F-117, built it to certain characteristics and specifications and told us what the radar cross sections were. We trusted and believed them, but we didn't know. We went out and flew the first night, and nobody was really in a hurry to go out again the next night. But the second night, they all came back. I went out the third night; we all came back. After about a week, we said this stealth stuff works. The first night, we did not know. The pilots did not know that certain people in the country's leadership said they thought we might lose up to half of our F-117s, but everybody at the base we took off out of were told that after we took off. The wing commander took all the senior leaders, and my wife was one of them, and told them. They would sit in the command post and listen as we checked in, as well. They made sure they heard from each of us that we were on our way home. Once we checked in, we still had about an hour and half flight left. Maybe after the first week, I think we all started to get comfortable that stealth worked. But we still knew going downtown, below 10,000 feet that at any time one of us could get hit.

Did the acceptable level of risk ever lower in your mission planning once you realized the low-observable technology worked?

No, we never changed anything. In our planning system we could tell where along our route we could possibly get seen and tracked. We tried to minimize that as much as could, and that's how we did the mission planning—the black line. The black line would have some time in it where we knew we could get seen or would be seen and could get tracked. However, we would minimize that as much as we can. We did that on every mission; we never changed. The difficult part of mission planning became as the days went on, we took out triple A sites, and we took out radar sites. We knew some of the threats were going down, but the intelligence community never wanted us to take anything out of our mission planning system. They wanted us always to assume it was still there. It was more difficult to plan, but we worked our way around it. You never knew when a site could come back up.

REFLECTIONS ON THE GULF WAR

How different do you think the air campaign in Iraq would have gone if we did not have or employ the F-117s?

The air campaign still would have been successful. In fact, the paper I wrote shortly after [at ACSC] was on how airpower could have won the war without the ground forces. It was not too greatly received by the Army, but it was certainly interesting. It would have ultimately taken longer as the stealth was able to get into targets that other aircraft could not get into easily and because we had the bunker busting GBU-27 bomb. We could go places where others could not. We were able to hit targets at the early stages of the war to take out the [integrated air defenses] so that the non-stealthy aircraft could get in and not worry about enemy fighters or SAMs being launched at them.

You also participated in and dropped the first bombs in support of Operation JUST CAUSE. How did that experience compare or contrast with your experiences in DESERT STORM?

JUST CAUSE was the longest combat mission I had ever flown. We took off out of Tonopah [AFB], flew all the way to Panama, and then returned home. The only reason the F-117 was used was that we demonstrated to the Special Forces at the Nellis ranges that we could hit an exact time over target and also hit where they wanted us to aim. Back in those days, they tried using the F-16s with limited success and F-15Es did not have their targeting pods yet. During JUST CAUSE, we did not even "stealth up" our aircraft. We did not even take the beacons off our airplane. The radar cross section was not at all minimized. Either way, there was a lot that went on in that mission, including bad weather. I had never flown on a tanker before below 2,000 feet. We greatly relied on that tanker since we did not have a radar altimeter to keep us safe from obstacles, and the turbulence that low is pretty excessive, especially when you are trying to take on fuel. But, the weather was such that we had to do that. In the target area, the weather was also poor causing me to employ the weapons below 3,000 feet. I had never employed below 5,000 feet before this. The humidity was also so high that the thermal returns received through the infrared system were extremely challenging and unlike anything any of us had seen before. So, that subsequently changed our tactics. We started practicing flying low along the California coast so we could see what high humidity would be like so not to be surprised if encountered again down the road.

During DESERT STORM, there were nights when the weather was bad. Sometimes they would give us high-priority targets and tell us to just do what you have to do to employ the bomb on the target. We would often have to go in so low that it would become uncomfortable. So yes, we did learn how to counteract this bad weather and fly low. With that, the lower you are, the less time you have to acquire the target. The bomb fall time is also shortened. Everything you do just happens faster!

How did the Gulf War experience shape or change you?

My experiences and most of others I've talked to is that the Air Force does a great job at training in a manner similar to that we will fight in. We practiced and practiced so that when we got into a combat environment, it would be natural. The hair would still stand up on the back of your neck as it is more serious than peacetime, but everything we did in peacetime carried over to combat. I am not just speaking for the F-117s, but for everyone that was over there. We learned a lot of things though, and our tactics, techniques, and procedures (TTPs) would change slightly as we continued to learn more. The fog of war and multitude of aircraft force that change. The missions in DESERT STORM each night were different than those of RED FLAG where you may plan it for 100 aircraft at most. Out there were many more, and each had to be deconflicted.

I'll give you an example of a lesson we learned. The GBU-27s could not be launched at the same time on one aircraft. If we wanted to put them in the same desired mean point of impact (DMPI), we would have to reattack. We learned though that once that bombs hit, the triple A started. They knew we were there. We also learned though that if we went out of the area for eight minutes, they would stop shooting. Therefore, whenever we would reattack, we made sure to go out for at least eight minutes to ensure that the triple A would stop before we arrived back in the target environment. We were not sure if they got tired of shooting or if their gun barrels got hot or they ran out of ammunition. Either way, we knew they would stop shooting, so we made sure to always go out for eight minutes when executing reattacks.

Another thing we learned a lot from was formation simultaneous attacks. We would do a lot of these deconflicted mainly by altitude to make sure the flight path of the bomb would not go through the flight path of any other aircraft. When we executed a simultaneous attack [anywhere between two to eight aircraft], we wanted to make sure we were on time because when the first bomb hit, the triple A would start. So, if any one person would arrive early, then everyone else would have to fly through that triple A to get into the target area, or if you were late, you would have to unnecessarily fly through that triple A instead of getting into the target area without getting shot at. We were very good with being precisely on time with our preplanned time over targets. More than one or two seconds off was rare, and if a guy got there early, he would hear it from the other pilots after returning to our home base.

Did you have a lot of RED FLAG experience before DESERT STORM and if so, did you get to practice with other dissimilar aircraft based on the sensitivity of the F-117 program?

The F-117 was declassified before DESERT SHIELD and DESERT STORM, but there had not been a lot of training with other aircraft. I did have experience in the F-15C at various RED FLAGs, and once we came out of the black program in the F-117, we did participate in RED FLAGs. The big mission planning cells did learn how to interact with us and fit us into the scenarios that they had there. But again, we mostly operated alone when employing ordnance. Adversarial red air [replicating an enemy] would still attempt to limit us, though. We quickly learned that they had a very difficult time trying to locate, track, and fire on the F-117. That was a great feeling.

What are your thoughts on the Air Force retiring the F-117 after a relatively short lifespan when compared to other platforms of the same era?

People often asked me, "What was your favorite aircraft to fly?" I always tell them that in peacetime it was the F-15C since all we did was dog fighting and other air-to-air missions, and that was a lot of fun! In combat, I said it was the F-117 because it was nice to be out there with only a slim chance of being shot down. We flew 1,271 sorties, and not one aircraft even got touched. As far as retiring the F-117, realize that it was 1970s technology. The B-2 came around with its 1980s technology and then the F-22 and its 1990s technology. This technology only got better, and you can tell that simply looking at the aircraft as they do not have the angles the F-117 had [along with all of their other advanced capabilities]. So, it was time to move on to something new. It was sad to see it go, but at least they did not turn it into a target drone like they do with many other aircraft in the Air Force inventory.

Additionally, some of the parts that made up the F-117 were no longer produced. The F-117 from idea to first flight was only 31 months, which is very rare. That was because they built it out of other aircraft. It had the ECS [Environmental Control System] of the C-130, the landing gear of the F-15, and the displays of both the F-15 and F-16s for example. That was how they were able to make it so fast, plus they did not have the same acquisitions oversight that we have today. The infrared system, both downward and forward looking, was getting old and they did not make them anymore. It would have been cost prohibitive to tear into the F-117 and replace them with something new. The aircraft was ready to be retired.

Now, when we got back from DESERT STORM, Secretary of the Air Force Don Rice came out to Nellis and had a few F-117 pilots sit down with him (I was one of the lucky ones) and detail what we thought would help improve the F-117 for the next 10-15 years. One of things we came up with was that we wanted an all-weather bomb, which led to the GPS Joint Direct Attack Munition (JDAM). We asked to be able to drop GBU-27s from a higher altitude than 10,000 feet. This led to new software so that during Allied Force, the F-117s were able to go ahead and drop GBU-27s from 20,000 feet, which would have been much more relaxing over Baghdad. Additionally, we wanted GPS for navigation since we only had inertial navigation system (INS) and we wanted to be able to drop both GBU-27s at the same time. Both requests were met. They upgraded the F-117 for as long as they could, but it did become cost prohibitive, and it was time to retire it.

What do you believe were the implications of DESERT STORM on the future of airpower?

It depends who you talk to. Again, my argument is that airpower could have probably won that war by itself. It may have taken a little longer, but I do think the coalition airpower proved that you didn't have to put boots on the ground. In my opinion, you put boots on the ground to clean up and stabilize everything after the fact. When was the last time that anyone from the ground was shot at from the air? Korea? It is pretty nice to have air dominance or air superiority where the ground troops can do anything they want without having to worry about overhead enemy airpower. So, I think airpower was proven in DESERT STORM in regards to what it could do. I think it could do more today though, and I don't think the Air Force gets credit for what it has been doing. Realize, the Air Force has been over in that area of responsibility (AOR) since DESERT STORM [Operation NORTHERN WATCH, Operation SOUTHERN WATCH, Operation ENDURING FREEDOM, Operation IRAQI FREEDOM] proving itself, and the public unfortunately does not understand this and neither does our Congress sometimes.

What do you think of most when you look back and reflect on your experiences during DESERT STORM? Any regrets?

The main thing that I reflect on from the DESERT STORM and DESERT SHIELD experience was the camaraderie of going to combat with your brothers in arms. I still have many friends I keep close contact with, and all we will ever do is trade war stories. No regrets!

LESSONS ON LEADERSHIP

You had the great fortune to return to the F-117 shortly after ACSC to command the 9th Fighter Squadron. Were there any lessons you carried forward from DESERT STORM to best care for the Airmen you were charged to lead?

Pilots cannot go to combat without the support it requires to get there. Be it the maintainers, logisticians, public affairs, everything is required, and my experience in DESERT STORM showed me how everyone plays together so that we could do what is required of us to fly the combat missions. Becoming a squadron commander, you are responsible for more of that team than you were previously. When I became a squadron commander, it was a time when the fighter squadron was also responsible for the maintainers. So, I had to learn a lot more about maintenance than I had ever known in the past. But again, you have to rely on your people. I had a chief and a first sergeant that taught me a lot, and you have to rely and trust those people because they are out there conducting that day-to-day business when you just can't be there.

Was the squadron you commanded rich in DESERT STORM combat experience or did you walk into a fairly inexperienced group of Airmen?

There had been a massive changeover in combat experience. Most of the DESERT STORM veterans had moved on to other things. I was one of the first pilots to come back as an operations officer and eventually a squadron commander. We were tasked to take on some more deployments, DESERT STORM and DELIBERATE FORCE for example, and my experience allowed me to relate to the squadron as they could better trust me. There were still a few veterans in the mix besides me. It was good to have them because they were the ones that most would lean on to ask the questions ranging from what do we do when we get there, what do we bring, what can we envision happening, and so on.

Were there any commanders, supervisors, or mentors from your past that contributed to your leadership style when you took command of your F-117 squadron?

All of them. Some good, some bad. I had a flight commander in the F-111 that if it were not for him, I would not even be a pilot. He helped me as a weapon systems officer (WSO) and as a navigator, and helped me write my application to get to pilot training. I had a squadron commander there that also helped me get to pilot training, and I learned a lot from both of them. Everywhere I've been, I've learned a lot from commanders, and I do not think I ever had any that were bad. Some were different. I learned things like how to delegate and trust your people. You cannot do it all, and if you try to do it all, you will drive yourself crazy. The people that you can trust will take care of you and you take care of them, and things just work out. The good thing about the Air Force is if you are working for someone that is not one of your favorites, you still learn from them, but you know in a two-year period that either you or they will move on and you will get somebody new. So, you can always learn from every experience you have and take the good and see the bad that you do not want to be like. If you didn't like it, the people that will work for you likely will not either.

What was the most memorable leadership challenge you have faced in your career?

I would say it was when I was the wing commander at Al Udeid Airbase, [Qatar], where I was responsible for managing many different countries and many different types of aircraft. I relied heavily on my commanders that worked for me, be it the group commanders that worked directly for me down to squadron and flight commanders. We once had a situation where one of the squadron commanders I needed to trust was not performing to the standards of mine or the Air Force's, and I had to relieve him of command. That was the hardest thing I had to do. We were in a combat environment, and when I relieved him of command, I had to send someone over to take his place. I also had to send the relieved commander home. It had to be done, but it was probably the most difficult challenge I have had in my career.

What were some of the thoughts that went through your mind when making that difficult decision?

I had come home to the States for a conference, but there had been a few issues before I had left with this individual. We had done some mentoring sessions together, as well had the group commander. When I was home, the group commander called me and said that he wanted to replace this squadron commander. I said okay we will discuss it, but not until I come back. I did not want to delegate relieving a commander to my group commander. Once I returned, I did my due diligence looking at all of the issues that were going on. Safety had become an issue, along with unprofessionalism in the squadron, and it was clear that I needed to relieve this commander.

You have to realize that we all have different personalities, every one of us. I'm convinced that there is not one person in the Air Force that intends to do a bad job or not perform up to standards. So, I think the first thing we have to do in any situation is talk to these individuals and make sure they truly understand the standards. We need to help them and mentor them before we take any action that will be detrimental to their careers. When you relieve someone of command, it affects the whole squadron, not just that one individual. There's also families involved. I think we owe it to them and ourselves to make sure we look into everything before making these life-changing decisions.

Did you have a personal relationship at all with the squadron commander you relieved, and was he receptive of your decision?

It's funny you ask. Yes, I had known this individual before from earlier in my career. He was junior to me, but I had known him while doing staff work. He was always an outstanding officer, but we all have to realize that not everyone is going to be a commander and not everyone *can* be [emphasis added] a commander. But, there is a role for everyone, and it is our job to make sure these folks get to these roles where they can excel.

Look, nobody wants to be relieved of command as they all think it will be the end of their career. It's not! In fact, this individual went on to make colonel. Deep down, he understood that he was in over his head and could not handle what was going on in this environment. To make things worse, this commander also lacked a strong operations officer to cover his back.

What was your leadership philosophy or guiding principles that helped you through this difficult leadership challenge?

I kind of already hinted at this, but you cannot do everything yourself. You have to be able to delegate and trust in your people. My philosophy has always been that I will trust them until they fail me. I owe that to everyone that works for me. Everybody truly is trying to do the best job they can. If they are not succeeding, you as the commander can either help them figure out how to do what is required to succeed or find them something else they can excel at. Like I said, everyone has a good intention. Nobody comes in and says "I'm going to do a bad job," and we must remember that.

How did you make sure that you continued to be relevant with the younger airman you would lead?

I would make sure to surround myself with them and learn. This crowd was not folks that looked just like me, either, or shared a similar birth year. Diversity is key, and you must listen to what everybody has to say. When I was the Air Force Chief of Safety, I started a program where there was one person from each base that had an issue with safety (motorcycle accident, not wearing a seatbelt, etc.) that would come out to the Air Force Safety Center where I would talk with them once every three months. They were typically younger in age and paramount in relaying issues of their counterparts from their respective bases. I would utilize these Airmen to come on the road with me to different bases and tell their story. Younger Airmen were typically more receptive, but older Airmen would learn from their mistakes, as well. They all had great stories to tell. None of them were crimes, though, simply mistakes. We are not in a one mistake Air Force, in my opinion. We are in a one crime Air Force, though. If you make an honest mistake, you should always be given a second opportunity. Allowing these Airmen to tell their story was my way to give them their second chance.

Thirteen years after DESERT STORM, you returned to Southwest Asia to lead the 379th Air Expeditionary Wing. What aspects of warfare did you find significantly changed from your earlier experiences in the same operating area?

Coalition warfighting had greatly advanced. DESERT STORM was the first time we had really fought as a coalition in the Air Force and it was pretty basic. Put simply, "Australia, here is your set of missions; Royal Air Force, focus on runways; US Navy, hit these targets; all of you, go forth and conquer." By the time I got back to Al Udeid during Operations IRAQI FREEDOM and ENDURING FREEDOM, we had advanced beyond that simplicity to the point that we planned airstrikes together as one package. Also, in the past, we had always interacted decently well with the Guard and Reserves forces. They flew the same aircraft and conducted the same missions, but the total force concept had greatly increased this since DESERT STORM. Additionally, technology also greatly changed. We were now operating with advanced targeting pods. Bottom line though was that the situation over there hadn't really changed too much for the Air Force. We still had absolute air dominance, and we provided that. There was no threat of being attacked from the air, and that makes things easier. Now, that doesn't mean that the troops on the ground were not in trouble, but the good news is we would always be available to provide over watch.

A unique challenge I faced when I got over there to command was that the Airborne Warning and Control System (AWACS) aircraft had redeployed home. That was the command and control method I knew from the past, but it would now happen through the Joint Surveillance Target Attack Radar System (JSTARS), Rivet Joint, and fighter jets remaining in the theater. This communications chain to the Combined Air Operations Center (CAOC) was a steep learning experience. Luckily, our Air Force is flexible, and we continued to do what it takes to get the mission accomplished.

How did the application of your leadership change from that of a squadron commander in peacetime to a wing commander at war?

I have always believed and said that it is your personality and leadership styles that got you to where you are. Do not change that as soon as you get a new job or a promotion. That is what got you there, and you do not want to make massive changes. However, you do have to think in a broader view. When I was 19th Air Force commander, I was in charge of 32,000 people. I was not able to meet each of these Airmen, just as I could not meet each of them when I commanded at Al Udeid. I had to rely and trust on my subordinate commanders that I would spend time with in meetings, meals, or wherever I could to ensure they understood my vision and my intent so it could be passed down to everyone. Subordinate commanders would usually pass along my words through commander's calls and then trust their capable chiefs, first sergeants, and supervisors for the execution of it. You cannot have anyone off doing their own thing, and one challenge of commanding is making sure everyone understands the mission and knows which way we want the wing to go. Maintaining open lines of communications, open door policies, etc. is critical to help ensure this.

Our Chief of Staff of the Air Force, General David Goldfein, is putting a renewed focus on leadership at the squadron level. With your experiences leading at many echelons in the Air Force, do you have any advice for today's squadron commanders?

I agree with what he is doing wholeheartedly and think it should have been done a long time ago. The emphasis has to come back down to the squadron commanders, and the squadron commanders have to be able to take on what we call the queep [unnecessary paperwork]. This is the stuff that doesn't really matter that adds hours to your duty day and may not even really need to be done. Yes, manpower is drawing down, though our new President and his senior leaders may change that, but some of the remaining manpower needs to be diverted towards the squadrons so those that are actually out doing the mission (e.g. loading weapons on aircraft, refueling jets, or ensuring the message to the public is getting out) have time to accomplish their primary job. We want them doing it well and not mediocre because they are tied up with additional duties.

What is one word you would use to describe the most important aspect of leadership? Why?

There really is no one word, but if I had to choose one it would be trust. Trust your people. They want to do a good job, and they will do a good job. You may just have to help vector them every once in a while. I think the commanders I saw that got into trouble were the ones that tried to do everything. There are only 24 hours in a day, and if you do not get sleep, at some point you will not be able to do anything. In DESERT STORM, I watched a commander try to do everything until he broke down and then could not do a single thing. So, you have to trust your people and delegate. There are many other words I could come up with, but trust is probably the most powerful.

If you could go back and give advice to Second Lieutenant Feest after reflecting on your career, what would it sound like?

Well, some things are just best left unsaid and unpublished from this interview! I'm very happy with my career. I would say not to change anything. Build relationships and make friends. Stay in touch with those friends because sooner or later you will regret losing touch with them. The key is to be the best at whatever job you have at that time and do not worry about anything else. Don't worry about filling squares and looking good because if you're great at your job, you will be taken care of. People will notice you. I honestly think that is what happened to me. I don't talk often about how I did as a WSO or a pilot, but I excelled as a WSO in the F-111 and that got me to pilot training. I did well in the F-15C and that what got me to the F-117 and leading the first package into DESERT STORM. When you find yourself in training, still do the best at that job, but also take some time to enjoy your family and friends. Shortly after training, you'll likely be deploying again.

What does it mean to you to be honored as an Eagle?

I never imagined I would be part of this! I often think to myself, why on earth am I doing this over others? I love doing this though because I get to talk to people and pass on some of my experiences. The part of this I really enjoy the most is when I'm asked questions and being able to answer using my wealth of experiences. If I wanted to tell my story, I'd write a book. I have no desire to do that, though it would probably be pretty good! [laughing] I enjoy passing down what I have learned throughout my career to others to hopefully help them as they go forward. It truly is an honor and a privilege to be selected as an Eagle.

Gen Feest revitalized America's confidence in its military with those first two bombs he dropped on the opening night of Operation DESERT STORM. He, along with his fellow F-117A Bandits, demonstrated to the world the awe-inspiring power of precision guided munitions in an environment rich with potential collateral damage. The US military demolished the Iraqi resistance with a short campaign that lauded minimal casualties, all thanks in part to the advantage that Gen Feest brought to the battlefield. His expertise at the tactical level carried over to his leadership styles as he always maintained his philosophy to "be the best at whatever job you have at that time and do not worry about anything else." This shaped his highly successful career, including leadership at the numbered air force echelon. The time spent with Gen Feest was an honor and the lessons we garnered will resonate with all and we thank him greatly for that!

■ Interviewed by Maj Will Powell, US Air Force, and
Maj David "Stevie" Underwood, US Air Force.

Lt Col Feest standing in front of his unit as the Commander of the 9th Fighter Squadron at Holloman AFB, New Mexico.

Maj Feest discussing lessons learned from Operation DESERT STORM with our 41st President, George H. W. Bush.

Maj Feest and his wife, Capt Bridget Feest, Tonopah's Logistics Readiness Officer.

353rd
FIGHTER SQUADRON
U.S.
U.S.

Paul T. "PJ" Johnson

A-10 PILOT/WEAPONS OFFICER | COMBAT SEARCH & RESCUE MISSION COMMANDER FOR DOWNED F-14

During the initial days of DESERT STORM, Major General Paul T. "PJ" Johnson heroically lead a bold search and rescue mission deep into enemy territory to locate a downed Navy crew. Born in 1958 in Gadsden, Alabama, Johnson spent his childhood in Tennessee. Prior to entering the Air Force, he worked on a hog farm and earned a degree in Agriculture from Murray State University, Kentucky. After receiving a commission from Officer Training School, Lackland AFB, Texas, he successfully completed Undergraduate Pilot Training at Laughlin AFB, Texas and was assigned to fly the A-10 Warthog. While serving in his first operational unit, the 353d Tactical Fighter Squadron (TFS) at Myrtle Beach AFB, South Carolina, Johnson demonstrated exceptional aviation skills and was selected to attend Fighter Weapons School at Nellis AFB, Nevada. When Iraq invaded Kuwait in 1990, his squadron deployed to King Fahd Royal Airport, Saudi Arabia, but Johnson's leadership directed him to continue his scheduled training. As a Weapon's School distinguished graduate and only days after completing his training, he rejoined his deployed unit. Five days after the beginning of the Iraqi air war, Johnson and his wingman launched on a mission to locate the crew of a downed Grumman F-14 Tomcat. Operating deeper inside enemy airspace than any A-10 before, Johnson successfully led the search and rescue effort that located the downed pilot. A few weeks later, he demonstrated his exceptional aviation skills when his aircraft was hit by enemy fire. Overcoming multiple system failures, Johnson managed to recover the badly damaged Warthog back to base. After the war, he applied his DESERT STORM experience as an instructor at the Fighter Weapons School. Johnson has served in staff positions at United States European Command and Headquarters United States Air Force. He also influenced professional leaders as an Air War College professor in the Department of Strategy and International Security. Johnson has flown missions in support of Operations DESERT STORM, NORTHERN WATCH, and ENDURING FREEDOM. He is a command pilot with more than 3,000 hours in the A-10 attack aircraft, and served as commander of the 75th Fighter Squadron, Pope AFB, North Carolina; the 354th Operations Group, Eielson AFB, Alaska; the 355th Fighter Wing, Davis-Monthan AFB, Arizona; and 451st Air Expeditionary Wing, Kandahar Airfield, Afghanistan. Following command, Johnson served as Deputy, United States Military Representative to the North Atlantic Treaty Organization (NATO) in Brussels, Belgium. Culminating his active duty career in 2016 and retiring as a Major General, he served as the Director of Operational Capability Requirements with the Deputy Chief of Staff for Strategic Plans and Requirements at Headquarters United States Air Force in the Pentagon. Among several other merits, the Air Force awarded Johnson with the Air Force Cross and the Defense Superior Service Medal. He currently resides in Henderson, Tennessee with his wife, Patricia.

EARLY YEARS

Growing up, did you find anything interesting that made you gravitate towards aviation or flying?

Ironically, growing up, flying was nowhere in my horizon nor was it in my plan, and it was something I never dreamed of doing. My intent was to go into farming.

In fact, the standard joke is that I went from "raising hogs to flying hogs." However, it wasn't until a retired Air National Guardsman from southwest Tennessee, Elliot Lambert, who spoke with me and encouraged me to consider the Air Force. Next thing I knew I was applying, accepted, and off to the races! I had never flown in an airplane until two months prior to flight screening at Officer Training School (OTS), and even though I never thought once of being a pilot, I did love to watch the crop dusters fly low across the fields when I was farming—pretty spectacular to watch. As they often say, timing is everything, and I joined the Air Force at the right time and towards the tail end of the [President Ronald] Regan boom. My OTS class was over 200 cadets, but the class after mine was only 60. Clearly, the Air Force was ramping down. I entered at the perfect time.

How did your experience growing up on a farm influence your career in the Air Force?

Believe it or not, there are a lot of similarities between flying a plane and driving a combine. It's all about unexpected noises. If something doesn't sound right the airplane talks to you and you pay attention to it. Paying attention to your environment is important. In farming, paying attention to the weather was important.

How did being raised in a small town influence your career?

The fact that it was a small community, everyone knew everyone, kept up with everyone, was nosey about everyone, but you had to get along with everyone or you would be ostracized. This translates to the military because the organization you are assigned to consists of people from many different backgrounds and cultures. We have to understand and learn everything we can about those people, so we can connect with them and overall work effectively with them to complete the mission. Coming into the Air Force, it opened me up to a whole broader world, but in pilot training, at least everyone there was experiencing the same stress, challenges and overall same shared experience that brings you closer together. When I was a junior in high school, I knew everyone in the high school. When I was in my squadron, I tried to get to know everyone in my unit.

Was there anyone who inspired you to join the Air Force?

Growing up I worked for a couple of men, one, in particular was Charlie Culver. I spent long hours farming with him, wonderful man who knew how to encourage and work well with young people, he was a natural at it. Also, Air National Guardsman Elliot Lambert. Elliot was a retired guardsman who spent 25 years working for various Guard units flying the original KC-97 Air Refueler, the C-124 [Douglas C-124 Globemaster II], and the C-121 [Lockheed C-121 Constellation]; he was such an encourager. Both Patricia and I had no family members or close friends in the Air Force so we were leery of joining and we knew once we did, it was going to be a radical move. However, Elliot and his wife Jane were both very encouraging and helped us through the transition. In many ways, my Air Force story is owed to Elliot Lambert, this is all Elliot. Without Elliot, the whole story doesn't happen. Elliot planted the seed, nurtured it and made sure it happened. Retirement coin #7 went to Elliot.

THE GULF WAR

Looking back towards the beginning of your Air Force career, how do you feel the jobs you held prepared you for DESERT STORM?

Very well, especially considering I was able to stay four years at Myrtle Beach [South Carolina] as a member, or "Panther" [Squadron mascot], of the 353 TFS. I wanted to go to weapons school, but the only way to prepare me for it and get me there was to make me the safety officer which shielded me from receiving an assignment, kept me in place longer, and got me the upgrades I needed to get me to weapons school. Keep in mind the late 1980s was a real Cold War presence. Our squadron's operational plan was to go to the desert if the Russians would invade the Middle East. Those were the plans we were prepared for and the part of the world we were training to go to, but always in the context of the Cold War. Our preparation was primary low altitude, and our intention was a low altitude fight, so I literally spent the majority of my time in the A-10 at 500 feet. Our deliveries were also low altitude deliveries. Fifteen, twenty, thirty-degree dive bomb deliveries, but when we got to DESERT STORM, we were releasing at 10,000 feet. We had never done the high altitude work, by our definition of high altitude. Although we were ready, there were significant tactics that needed to be adjusted before we went. I had been selected for weapons school in the fall of 1990, Saddam invaded in August, and the jets were configured to go. I had to depart for weapons school the day my squadron left for Saudi Arabia—I was miserable. Not only did I miss the opportunity to deploy with my squadron, but I also missed the birth of our first-born.

While at Weapons School, was there any training you could integrate back to your squadron at the time?

No, not much because at that time the A-10s were there as a speed bump in case Saddam came south to help forestall the move south. The airplanes remained on alert with only a half tank of gas and limited ammunition. Reality was changing for the squadron while they were there and as the war began to evolve. The Panthers were also the first squadron to arrive and by the end of the war, there was a total of six fighter squadrons participating. It's hard to fathom how difficult communication was during that time. If you were able to get to the morale tent for a 10 minute phone call home, then you were doing well. There was no email, so you sent a letter. The atmosphere at the weapons school was to ignore what was going on in Saudi Arabia and concentrate on the syllabus.

Cold War training mentality at weapons school? Was it difficult to transition to DESERT STORM?

Altitude was much different as we were flying in the teens in the desert, and at home we flew much lower. The airplane was loaded down and was very heavy, we were scattering ordnance throughout the day at high altitudes which really was the only adjustment for our day fliers. Those that flew at night had a radical change. They used missile technology attack weapons and used it as a flare, lights out, no illumination of any kind, night tactics at 5,000 feet—dramatic tactic changes. The weapons school maintained current tactics syllabus until they had lessons learned to apply. In the fall of 1990, a major decision was made announcing the war would be fought high. Navy fought low the first night and couldn't believe they all made it back safe, they flew high for the remaining time in the Gulf. Bottom line, the old European war mentality had to be modified and people adjusted well to it.

How quick was the transition from weapons school to joining your squadron in Saudi Arabia?

Very quick transition. The evening I graduated weapons school was also the same evening I received a letter from my squadron commander to join the squadron in Saudi Arabia as soon as possible. I graduated and spent two nights at home, flew back to Myrtle Beach to get deployment gear, and got to Dover AFB [Delaware], awaiting a C-5 [Galaxy] for transport. Despite nearly two weeks of multiple aircraft maintenance delays, I finally made it to the AOR. I arrived in country two weeks before the war started and took teasing from the squadron for living in Vegas and letting time go by just up until the war began. Squadron teammates were there for several weeks working and waiting for things to kick off. I knew everyone very well once I arrived in country and assumed the weapons officer position from [Captain Steve Phillis] "Syph" who was our only Killed in Action (KIA). We were a pretty experienced squadron in terms of flying. The least experienced guy was [Lieutenant] Rob ["Sweetness"] Sweet, Sweetness had his mission ready check ride complete a week before the unit deployed. However, he did have one unaccomplished task (night air refueling), so Sweetness was paired with Syph during the flight over and they were always paired together when flying, they were also the team shot down and Syph was the Squadron's only KIA. Sweetness was shot down and became a POW; Syph flew low trying to cover for him and was KIA. He was a natural born weapons instructor and the ultimate weapons instructor and losing him was difficult for the squadron.

While deployed, what did you do for fun, if time allowed?

Most of the fun occurred before I got there and before the war kicked off. Once the war started, you were either in the jet, in the rack, or in the dining hall. Near the module "hooches" where they were constructing King Fad AB, they built a cold tub piled with sand bags, lined with tarps filled with water. Most of the boredom occurred prior to the war, and morale was at its worst during November due to the fact that there were so many unknowns. During the war, there was rarely a day off, airplanes were flying 4-5 times a day, mission capable ready airplanes were 95 percent or better, young airplanes with plenty of parts and I have never seen such a motivated group of people like I did with that group of young maintainers...it was just incredible.

What was your mentality prior to stepping to the first mission? Where you scared or excited before it all started?

We in the A-10 were unsure of what we were going to do. We are CAS, but there was no ground fight at the time. As of day one, we were going to start with the kill boxes. Some of us, however, did a different mission on night one. Air Force helicopters escorted in F-117 Nighthawks and F-15E Strike Eagles to take out the EWR network. They successfully blew a gap in it and they wanted to continue to make that gap bigger. I love telling people my first combat mission in the A-10 was not CAS, but offensive counter air (OCA). I did OCA in an A-10. [Naval Lieutenant] Larry Butler on day one contributed to taking down the EWR system. I was rolling in on my first pass, using G-model 300-pound maverick precision, air-to-ground missiles, I reached over to slew and realized the entire slew button assembly was disassembled on the floor. At this point, the apparatus consisted only of a sharp point at which I had no choice but to make work. Of all times—day one, target one, and this happens? On that day, we hear on the radio from the AWACS [Boeing E-3 (Sentry) Airborne Early Warning and Control aircraft system (AWACS)], that [Iraqi MiG-29] Fulcrums were heading our way. However, they disengaged and turned back immediately upon the Strike Eagles engaging them. I was pumped, anxious, and overwhelmed before this all kicked off. When we pulled out of King Fad, the level of activity at King Fad was unbelievable, airplanes were everywhere! Everybody was loaded. We were loading and refueling with engines on, turning aircraft in less than 40-45 minutes. As I was taxing down to get fuel, maintenance, and weapons, personnel were holding cardboard signs that read, "Weapons here, cheap!" They wanted you to get your weapons from them [with emotion in his voice and tears in his eyes]; it was neat to see. Keep in mind, we were 19 years post-Vietnam. Only the wing, vice, group and squadron commanders had combat time—that's it! When I went through pilot training, I flew with one instructor who had combat training. You go to A-10 training today and every instructor has combat experience. It was hard to fathom that only people who have been there with the experience were your first-line supervisor, squadron, and wing leadership.

Five days into the war, you're tasked with a search and rescue mission. Can you tell us about that day?

I flew every day leading up to the CSAR mission. CSAR was an additional mission conducted by the A-10. Each squadron has approximately six to eight "SANDY" qualified [CSAR qualified] crews. The SANDY burden did add required sorties and missions, and it was challenging to get it all accomplished. I was one of them and everyone wanted that mission—they had the best stories. Air Force Special Operations Command (AFSOC) would do the pick-up, their method is to use coordinates and go in at night; [MH-53J Special Operations] Pave Lows would go in and do a night pick up. The Joint Forces Air Component Commander (JFACC), General Chuck Horner, ordered us to conduct CSAR, but we didn't have a lot of time to practice with the Pave Lows. The irony is everyone wanted to be a SANDY pilot, but no one wanted to do it. First, no one likes to sit alert, if we have to get up at 0430, let me fly, if I'm not going to fly, let me stay in the rack. [Captain] Randy "Golfball" Goff and I were sitting alert and our plan was to fly up to our forward operating location (FOL) and sit at that location because it's closer to the fight. En route to the FOL, we call up to Operations for a weather check and we were told to stand by for a tasking. Not used to ever conducting a CSAR tasking, we were confused on what type of tasking. It was a CSAR tasking. So, we were off to the races. We receive the coordinates which were way out west. We think the coordinates came from a [Boeing RC-135 reconnaissance] Rivet Joint off a survival radio. During this mission, we were in coordination with the downed F-14 aircraft pilot, [Navy Lieutenant] Devin Jones and his radar intercept officer, Larry Butler, the helicopter pilot, [Captain] Tom Trask, and my wingman, Golfball. The weather was bad in Iraq, but that didn't stop Tom Trask who was up at 100 feet with 0-0 visibility doing radio search. At 100 feet they were unable to get radio line of search and had to peel out and come back to get gas. Meanwhile, we got diverted to a SCUD [tactical ballistic missile] hunt; this is how important the SCUD hunts were. Everyone was obsessed over the SCUDs. As we approached the area, we could see at 15/18 [thousand] feet what they were talking about, the proportions looked right. However, at the time, I did not have assets, we were loaded for a SANDY mission. So we requested a fighter who was prepared and equipped to drop bombs. His first attempt was a clean miss. So, I went down below 10,000 feet to get a good look and I see the goats run. This particular tribe configured their goat tents that appeared to look like SCUDs. No one was hurt. Time for gas and back to the CSAR mission. In an attempt to locate the tanker, we realized we were carrying the same last three numbers (mode three) as someone else, and as a result, we were getting bad vectors. Bad weather and bad vectors, but finally we pop through a cell and can get gas. We can now start our search. Pressing with the coordinates we were given, the weather starts to clear, and we approach Iraqi territory. No GPS and my INS was already airborne 3-4 hours (single INS) is drifting—somewhere in Iraq. So, I start calling for Devin and Larry who were conducting an escort mission for a Navy E-2 [E-2C Hawkeye]. During their mission, their F-14 Tomcat got too low and was shot down. It was late at night, with high winds, and Devin and Larry got widely separated during their ejection, so once they got on the ground, neither one of them could make radio contact with the other.

Radio signals were not good. I had to fly in several directions west and north to locate a strong signal until I finally contacted Devin. At that point, I was able to finally start searching on the ground for him. Terrain made it difficult, with no trees or mountains to distinguish from. I had to duck below the clouds to get a steer directly over them but at this point I'm now low on gas. I requested a tanker that came into Iraq. Tom Trask meanwhile is listening to all radio traffic and requested the coordinates. He takes off towards the coordinates, which are INS coordinates that were not accurate. With bad coordinates, I came below the clouds and completed a daisy chain and continued this revolving escort for the helicopter. Approaching the coordinates, Tom finds a gap above the heavily congested four-lane highway and pushes onward towards Devin. As we approach, we spotted a truck headed straight towards Devin. I roll in and fire with a clean miss. I pull off and Randy rolls in and catches part of the truck, confusing and scaring the driver. I roll in on a second approach and smoke the truck. At that time, I was unsure if it was a military truck. Randy was convinced it was not and thought it was just a farmer's water truck. We weren't sure. But, we had a survivor on the ground, six people in a helicopter, and the truck needed to go away. So, we fly towards the truck to find Devin and shortly beyond the truck Devin stands up and comes running towards the truck. Here's the irony, we used almost the same exact tactics as used in Southeast Asia [hover cover, daisy chain, etc.]. We then continue to escort south to the border and heard radio chat of possible Iraqi air defenses (red air). At that point, I requested an Eagle Combat Air Patrol to provide cover and as additional support moved in, radio chat became busy and out of control. I ordered everyone but Tom and Golfball off the radio. For the fourth time, I needed more gas (NOTE: Three of four refuels came from the same KC-10 [Douglas KC-10 air refueling aircraft]). We return to King Fad AB after an 8.8 hour sortie. Tom took Devin back to an airstrip out west and in three days, Devin was back on the boat and out flying missions. Larry was captured and got repatriated. When he was captured, he had no idea what happened to his front-seater.

On a side note, the first time I ever met Devin was at ACSC. Devin was a student and I was there as an Eagle invitee telling this story. I was invited to Eagles but had no idea Devin was there. Attending the banquet, I sensed space behind me and was tapped on the shoulder and there stood Devin Jones. It was the first time we spoke since we talked on the radio that day.

Looking back at that day, is there anything you would do different?

Yes, tactically, I blew too much of the Isolated Personnel Report (ISOPREP; personal recovery details required) data too soon. Devin was one cool cucumber on the radio, so much that I was suspicious and I ended up blowing two identifiers right away. I needed to rely more on my wingman Golfball to run back up on the radio/SANDY-2 function. However, Golfball's radios were poor due to weather. I should have tried later in the day to change up the radios and had Golfball monitor that. I may have prevented the radios from getting muddied. I spent too much time on the radio with Devin because we were convinced they would automatic direction finder (ADF) his radio.

A year later, I'm at Nellis AFB, Nevada as an instructor conducting CSAR training and we are working with the Pave Lows out of Kirkland [AFB, New Mexico]. As I was talking with one of their flight engineers about this story, he asked about the location and where it happened, he also asked about the truck and I told him we didn't know for sure what it was. He said, "We found that truck a few weeks later. We were driving across the desert and out of nowhere, absolute nowhere, is a blackened Iraqi truck. So we investigated and found 10 dead Iraqi soldiers in the back." So, we were convinced they were ADFing on Devin's signal. We were probably 60 seconds from being too late, which made me realize I spent too much time on the radio with Devin.

You had experience with aircraft battle damage, can you share what happened that day?

As a Liaison Officer (LNO) at Eskan Village, Riyadh, Saudi Arabia, we came up with the war [F4G] Weasel plan. Iraqis would turn off their SAMs and therefore F4G Weasels equipped with radar-seeking missiles and tasked with destroying the SAM sites, were unable to take them out. They could suppress them, but not destroy them.

The idea was developed to take the [General Dynamics F-111] Aardvarks, A-10s, and F4G Weasels and perform Destruction of Enemy Air Defenses (DEAD) to take out the Iraqi SAMs. Reluctant to do it, we had to do the mission, and as a result, we successfully destroyed an SA-3 site—plan worked. A-10s are now going after SA-2/3 sites. Once I was finished serving as an LNO, I went back into the fight and was tasked to take out an SA-2 site. Visibility on the target was not good. I made the bad decision to throw some bombs on it to see if we would get any results. By just dropping bombs on the site, I couldn't verify what I did. I should not have done that. The weather got worse, and I lost sight of the target. I missed my wingman's radio call, so no one was looking at me. I'm not in position to hit the target, I had bad perimeters, but I shot at it and at approximately 6,000 feet I'm low, and as I pulled up and started climbing to 7,000 feet my right wing took a hit from what we think was a shoulder-fired SAM. The airplane hard rolls right and for about 10 seconds I thought I was about to eject. As I'm about to reach for the handles, the airplane rights itself and I slowly leveled and limped out. When the warhead came through the wing, it blew debris into the number two engine which stalled the motor. Although the hit foddered the engine, the engine spooled back up, and I had just enough thrust on it to get wings level and get the airplane back home. Not one of my star moments. Ninety miles out on final approach I realized I needed gas. My fuel was hit at impact and my fuel gages were not working, I wasn't sure how much gas I had or needed. I requested a tanker who at the time was sitting at 29,000 feet. It slices through the stacks to refuel me. Limping along at 200 knots, the KC-10 slides up on the right, rolls the flaps, and throws down the gear to slow down to match my speed and we are able to refuel. Refueling speed was at a terrifying slow 170/180 knots with no hydraulics, I was unable to latch to the boom. I successfully got gas and landed. Upon landing the right tire shredded on landing, and various fluids were running from the aircraft but a successful landing overall. It was monumental. Irony was she was my airplane, tail number 2664. My crew chief was not pleased. Sergeant Eric Stanaland fixed the aircraft and got me and the airplane home and back to Myrtle Beach. Ten years later when I took command of the 75th Fighter Squadron, 2664 is present and on display. She's a member of the family. She's in the bone yard now and an organ doner for today's active A-10s. We even went to see her in Tucson, Arizona.

Prior to getting shot at, did you have any other experiences with this or only in training?

Only in training. I've been very fortunate in the A-10 with the number of engine failures and gun jams. What was remarkable was in 30 days the maintenance professionals and engineers put 2664 back together with several new parts. When we were getting ready to head home, the only thing you could see sticking out of the canopy of 2664 were Eric Stanaland's feet. When he was asked if he was ready to go home, he stated [with emotion and tears in his eyes], "Not until my airplane is fixed." Everyone is a volunteer. It is what we have now, and it is what we have had for the last 15 years. People have all deployed several times and the sacrifice they have given and the time spent away from their families, and over time, I'm not the least bit worried about the people we have fighting in our Air Force today. I'll put them up against anybody.

How has your experience in the Gulf War helped you in the jobs that followed after the war?

We (not I, we) put CSAR back into the weapons school syllabus. Those of us at Nellis AFB post-DESERT STORM, advanced CSAR, and incorporated night CSAR training as well. We'd like to think we contributed to enhancing the CSAR community. Also, lessons that you learned that you wish to pass on is that there were guys that did stupid stuff every day and came out without a scratch. There are guys who did it right every day, and it didn't matter...and that was Syph. I was in the middle, I did something stupid, paid the price for it, but nobody got hurt. So how do you teach somebody else about that? When I discuss that lesson, I admit that I took that target personally. Growing one eyebrow and a set of fangs, I was determined to take out that target. There is also no room for hoo-uh in the cockpit, there is room for it in other places, but when you are flying aircraft, there is no room for hoo-uh in the cockpit. You always have to be in control of your intellect, the process, mentally ahead of the airplane, mentally ahead of the game and you can't allow yourself to get to a "hoo-rah/hoo-uh" mentality in the airplane and that is where I got! A far different experience years later as a colonel group commander at Bagram AB, Afghanistan. As a wingman, I'm flying right on the Pakistan border looking for targets on the ground. As I came out of a turn, the number two engine quits. I handled this situation differently, knowing I had time to analyze the problem, and knowing I didn't need to be in a rush, in fact, very few times do you need to be in a hurry in the airplane, never take it personally, and have radio discipline. Never key the mic unless you have full situational awareness. Never let them hear emotion on the radio and never let them hear your attitude on the radio because that drives your actions in the plane. I let that get away from me in DESERT STORM. The way we train is so effective, the emphasis we put on it and the way we make it as realistic as possible, ensuring our people are ready to go from day one. I like to think our generation helped contribute to that effort as best we could.

How did the relationships you built with those you worked with down range help you communicate with the different platforms? Are you still in touch with some of those people?

I've ended up as an instructor pilot with some of these folks, Intermediate Developmental Education at Leavenworth, and time on staff together. Some of our closest friends have been established since the day before DESERT STORM started. You do make those relationships that matter. What we had to get better at after DESERT STORM, especially in the A-10 community was composite operations. We were not always good at integrating with a larger force, and I really learned this as an instructor pilot at the weapons school. In today's world, everything we do is composite operations, Kandahar [Afghanistan] alone had nine different weapons systems, and everyone was working with everybody else. In DESERT STORM, we were far more stove-piped than we are today, and we have climbed out of that stovepipe to some extent. That's a good thing and we can always do a better job at it, but we are better than we were back then.

Did you encounter any Total Force Integration (TFI) concerns when you tried to incorporate the Guard and Reserve into operations?

Downrange, no. Downrange if I can tell you're a reserve component, then something is wrong. It was a rare exception that I could do that. The challenge is really home station because the duty status is so different. The reserve component brings such vast experience, and the active duty component brings availability. As a downrange wing commander, it was transparent; the real challenge is home station.

FORGING LEADERS

How has this experience shaped you as a leader?

First, remember that frequently not always, but frequently, if something happens worthy of a medal, it is because something went bad to begin with. A guy got shot down and talking with Devin during the 1997 Eagles, I found out he was not in the auditorium when I did my interview. We had a nice chat prior to my presentation, and he told me he was not going to be there. He went on to say it was a great story for me, but it was a pretty painful memory for him. He got shot down, and his backseater became a POW. It wasn't a good memory for him, but it sort of changed the trajectory of my career. We often joke, "Timing is everything and there is no justice." Well, timing is everything it was my turn to sit CSAR alert that day. I have absolutely no question in my mind that any SANDY qualified pilot would have successfully completed that mission, I just happen to be that guy on that day. When good things happen to you during the course of your career, be thankful, it could have happened to someone else. When unfortunate things happen during your career, it may not be because you did anything wrong, it just simply happened. If I would have applied to OTS six months later, I would not have competed or made it. I don't have a lot of patience for those people who feel they earned it, to include those people who think they earned the right to lead, command or get promoted. I have been incredibly blessed to have worked for incredible bosses that gave me opportunity and nudged me in the right direction. Tom Trask and I sat down and tried to count how many individuals were involved in the CSAR mission; we came up with approximately 35 people—a lot of people—and no one said no. I asked for a tanker to come north, and they came. I asked for an Eagle Combat Air Patrol (CAP), and they came. Nobody said no, and that says a lot about our heritage in our United States Air Force. I was honored to be a part of that CSAR mission. As a leader, it shaped me in many ways, and at the same time, I believe life experiences shape you for leadership experiences. The key is to pay attention, listen, and learn from everybody else. Not all lessons are good lessons, but every lesson is an experience.

What is one word you would use to describe the most important aspect of leadership?

Servant leadership. When I was a wing commander, I really enjoyed leadership when I could sit back and watch Squadron Commanders lead their squadrons. That was fun! When I saw Squadron Commanders take care of their squadron and then see how their squadron would respond, was truly neat to see. I could also see those who were in it for themselves, and you see how the squadron reacted to that. We are the profession of arms that really is something bigger than ourselves. What is wonderful to watch is when you see a young Airman get that. They don't get it initially, but at some point, they get it. And they realize this is bigger than me. Some chief, some SNCO [Senior Non-Commissioned Officer] or supervisor at one point helped Sergeant Stanaland see it, and get it, and he realized he wasn't ready to go home until his airplane was ready.

What's been your greatest lesson learned as a commander/leader?

My general philosophy as a commander is to try and be the coach. Try to show people the end zone and get out of their way. Ninety-five percent of the people want to see the unit succeed, let them do it and be willing to let them do it not your way. Let go of the reins and let them run.

What was your most memorable leadership challenge and how did you overcome it?

Every organization has a culture. If you are new to the organization, you need to learn what it is and figure it out. The second hardest thing is changing it if it needs to be changed. Be aware of the trap. Sometimes the culture does not need changing and know when not to monkey with a culture. It's hard to change a culture that needs changing especially in the short amount of time one is serving as a commander. You have to be okay with the fact that not everyone is going to like you changing the culture, but you can't be afraid to change it if it needs changing. However, understand exactly what it is you are trying to change and why. Sometimes you only have enough time to understand the culture you are in and learning how to deal with it.

Are there any leadership risks that you took or wished you would have taken?

I had to let go of two commanders; one was clear as to why he had to go, and the other was due to losing confidence in their ability to lead. The latter was very difficult. There are times when you question if you did the right thing or not. However, there will also be times when you will need to make a decision. You make the decision, you move on, and you accept the consequences of that decision. We have all seen leadership styles where our bosses could not make a decision. They want to wait until they have all the information to make that decision and by that time, it may be too late. You have to find some level of comfort with ambiguity to make a decision with less than perfect information. Your hardest decisions will be personnel decisions and disciplinary decisions; do I take money, stripes? It's important to keep in mind you are making a decision to benefit the whole, not just that one person.

Looking back at your tenure, is there anything you know now that you wish you could have told yourself back when you first started?

The one thing I would change was when I was back in pilot training. I was a nervous wreck in tweets [Cessna T-37], I was so nervous. Learn to just enjoy yourself. I recall my T-38 [Talon] instructor pilot singing in the back seat. All he was teaching me to do was to have fun. Whatever you do, try to enjoy what you are doing. Enjoy the people you are around, enjoy the job you got and remember, if you don't like the job you're doing, stick around a couple of years, and you'll do a different job. But really, this is an interesting question you pose—I don't think I would change anything.

Although Johnson may have had the opportunity to continue to serve and advance, in 2016 he chose to retire. He "knew it was time." With his wife Patricia, they are looking forward to spending more time with their family, supporting their community, and enjoying the next chapter of life together.

- Interviewed by Maj Erika Chute, US Air Force, and Maj Jessica Guarini, US Air Force.

Paul Johnson rides in a parade car with Brooke Shields during a DESERT STORM welcome home celebration.

Capt Johnson stands in front of the A-10 that brought him home a "on a wing and a prayer."

Paul Johnson sets the standard as an A-10 weapons instructor.

Paul Johnson with his wife, Patricia.

Guy Hunter, Jr.

OV-10 AERIAL OBSERVER | DESERT STORM PRISONER OF WAR

CWO5 (Ret) Guy Hunter, Jr. grew up on a farm in Moultrie, Georgia. The son of a South Georgia farmer, Hunter enlisted in the United States Marine Corps after graduating from high school and following in the footsteps of family members who had served in WWII. He went through Marine Corps Basic Training at Parris Island, South Carolina in 1962, and the next year was assigned as a radioman in Vietnam for the 1st Marine Air Wing Pathfinder platoon. Hunter spent four tours in Vietnam, performing multiple duties, culminating in his assignment as a Naval Gunfire Spotter. He gained valuable on-the-job training in Aerial Observer missions in the Thomas-Morse O-19 and the O-1 Bird Dog. Hunter became a full-time Aerial Observer, crisscrossing between coasts in the United States, and was eventually assigned to Marine Observation Squadron 2, the unit he deployed to Iraq with during the Gulf War. In November 1990, Hunter's Squadron, Marine Observation Squadron (VMO) VMO-2, received the word to deploy to Saudi Arabia in support of Operation DESERT STORM. On the morning of January 18, 1991, Hunter's OV-10 Bronco, piloted by Lieutenant Colonel Cliff Acree, was shot down by a SAM in Kuwait. Injured from the ejection, Hunter and Acree were unable to evade and were captured by an Iraqi unit in the area. Both Hunter and Acree would spend 47 days in captivity, and at the age of 46, Hunter was the oldest Prisoner of War (POW) during all of Operation DESERT STORM. Malnourished, tortured, and kept in isolation for the majority of his captivity, Hunter remained committed to returning home with honor to the country he swore to protect. On the night of 24 February 1991, Hunter's cell was rattled by an explosion. An F-117 Nighthawk bombed the prison facility in Baghdad, where Hunter and fellow POWs were being held. The bombing was so intense that it blew the transit windows out of the walls and buckled the prison doors. No POWs were injured in the bombing and they were relocated to another detention facility. During his captivity, his wife, Mary, never lost hope and on 5 March 1991, the two were united when Hunter and his fellow POWs were released. Hunter would go on to serve three more years in VMO squadrons until the OV-10 Bronco was retired in 1995. At the time, Hunter was the longest serving and most experienced Aerial Observer in the US Marine Corps. CWO Guy Hunter retired with thirty-two years of distinguished service in the US Marine Corps. He currently lives in Jacksonville, North Carolina with his wife Mary.

PRE-MILITARY LIFE

Would you mind talking about your childhood—where you grew up, when you were born, and what childhood was like for you?

I grew up on a farm in Moultrie, Georgia. My parents and grandparents owned a 440-acre farm so we were continuously busy. We grew watermelons, cantaloupes, tobacco, tomatoes, cotton, corn, peanuts; we had a couple of pecan orchards and a couple of pear orchards. You can imagine we were busy all the time doing things. When I was a small boy in south Georgia, those areas were still in the last throes of the Great Depression. They hadn't come out of it just yet. What we used for power on the farm were mules. We had twelve mules on that farm. That's what we used for plowing and dragging the rakes for gathering the hay and the peanuts, and also for pulling the wagons around. I taught math at a community college for a few years after I got out of the Marine Corps. I used to tell my students that when I was in elementary school, I think it was like third or fourth grade, perhaps. I told them that when I got out of school, I had to take off my shoes because we weren't allowed to wear our school shoes walking around on the farm due to the dirt. I had to go hitch up a mule to the plow and go out and plow until it was dark. Then I would go back to the barn, undo the mule and harness and everything, put the mule back into the mule yard, go out to the well, prime the pump, put water into the kettle, then I had to go into the barn up in the loft and throw hay and corn down into the haymow, a little v-shaped building you know, so the mules could stick their heads through and eat the hay and corn. I used to tell them [the students] that's what we did back and the day and they looked at me like I was from Mars or something—they just did not understand that kind of life.

It was fun. On the weekends, the boys in the neighborhood would get together and run around through the woods, build little houses and forts, camp out, and go hunting and fishing. In fact, we hunted all the time. We lived in a nice sized house with a hallway through the middle. There was no air conditioning back then, just huge, high roofs. And so, all the windows were half-mast in the summer, so we always had good ventilation. I just don't recall it ever being hot in that house. Along the hallway going through that house were all the rifles and shotguns up on the racks. If we weren't doing anything on the weekend, we would take one off the wall and head out with some bullets or some shotgun shells. That's how casual we were around guns in those days. When I went to high school, all the boys had shotguns or rifles in the back of their pickup trucks they were driving to school. It was considered the "norm."

What made you want to join the military?

Always wanted to. Yeah, my uncles were in WWII, you know. One of them was there for D-Day. In fact, there's a little article in the paper in the Moultrie Observer several months ago, about my uncle Randall, and they found out he had some belated medals—a bronze star and others, for doing the landing in WWII. My other uncles were in the Carrier Navy in WWII. So, I've always wanted to do that.

Did you make the decision early on in high school, when I get out I'm going to go directly to the Marine Corps?

Yeah, I did that. I joined the Marine Corps before I got out of high school. The Marine Corps had something called the delayed entry program where you could join up three or four months prior to going to boot camp, you know. You didn't have to do anything but you were a member of the Reserves Service, but you didn't get paid for it, but it built time in service and time in grade for me, for several months already. So, I went to boot camp; I got my orders and went to boot camp in July. I checked in to boot camp and was still seventeen years old.

What year was that?

In '62. The next year in '63 I found myself in Vietnam the first time.

Were there any other kids in your high school that also joined the service?

Later on, they did. My best friend [Lynwood Murphy] went to North Georgia Military Academy. Now it's North Georgia State College. In those days, it was North Georgia Military Academy. If you graduated, you were guaranteed a commission in one of the four services. It was a four-year college. So Lynwood joined; I guess he got bored with it, and he quit school. He enlisted in the Marine Corps and, in '65 I was ashore with the 4th Marine Regiment and I was walking around chasing the "communications line," they were always broke where we were set up, and there was Lynwood sitting in the bushes. He had just come ashore with the 7th Marine Battalion. We were getting ready to kick off this big operation. We chit-chatted a while and talked about what we were going to do when we got back, you know, because we were good buddies. We grew up right side by side on the road, went to school together, rode the school buses, dated the same girls, went camping and hunting together, fishing together, so I was happy to see him. A few days later, he was there when Operation STARLIGHT started, the first big Marine Corps operation over there. I was a naval gunfire spotter by then. We ended up killing about eight hundred of them, and after the operation started, about three days later, Lynwood got shot in the chest by a machine gunner and died there on the spot. His Marine Corps time was about a year and a half and his time in Vietnam was about six weeks. I tell people all the time, this stuff about the good guys and the bad guys, that's movies… that's not real life. In real life, the really best guy in the world can get shot the first day and never make it to real combat, or the really good guy could make it all the way through and go home safe and sound. The really bad guys, same thing. They could get waxed the first day, or they might do something or work it out where they slide through and slide through and get through combat without a scratch on 'em. There's no rhyme or reason as to who gets killed in combat. It's just where the bullets fly.

MILITARY SERVICE

What was the atmosphere like when you joined the Marine Corps, both in the United States and in the service?

I thought everyone was pretty patriotic. I thought there was good support for the military. I know a lot of people don't like to hear this, but every time you hear some of these guys talking about "we came back to the States, and they were spitting us and calling us baby killers," you know I think that may have happened like one time. You ever see the Seinfeld episode where he talked about getting spit on? I just think that's one hell of a spit because I was back and forth more than any of those guys in the States, and never saw that ever happen. And if it had, I would have "cleaned his clock" because that's assault. I could defend myself then. I wouldn't have put up with that for a second. I just don't believe it happened like that; I think it's just a myth. People treated me nice, you know. I came back once—flew into Travis AFB [in Northern California], and didn't have any orders yet so they put me up in Treasure Island for a couple days until I got orders. They finally got me some orders to go back to [Camp] Lejeune [near the coast of North Carolina], and I went to the international airport in San Francisco and got on there and they said, "We have an empty seat for you up in first class, come on." They all treated me very nice. I was treated like a very distinguished person; I was a Corporal at the time. I never saw anybody, any military man treated badly once in my life. So, all I can say is I think a lot of that is myth or just pure wanting people to feel sorry for them.

What were things you loved and disliked about the Marine Corps when you joined?

At the time, I didn't like, you know, filling sandbags a lot. I didn't like being on the work parties. But, by the time I got to be corporal and sergeant, I said, "You fill the sandbags." So, that feeling went away. But I enjoyed it all, you know. I liked the structure; I liked what we were trying to do, always training toward some goal, you know. To get better. When I was very young in the Marine Corps and putting up communications, the big radio antennas, getting all the radios remoted into the fire direction center, that kind of thing, and doing good at it. It was fun too, seeing some of the officers there because, you know, like our fire direction officer was the operations officer for the battalion and he'd look back there along the back row where all the guys were sitting with their radios for calls for fire. If he saw you nodding off and taking a rest, he'd knock the snot out of you. Some of it was just flat funny. He made us all laugh sometimes when he'd start getting on some of the troops back there. He wasn't being malicious, just being funny and making 'em wake up. It was just great.

But promotions were dead slow back then in the early '60s when I came in. I'll tell you one example, I was up in, when I got out of boot camp, I went to Marine Barracks up in Norfolk [Virginia] for a while, and in those days, Marine Barracks had these huge brick buildings and inside the buildings was the quarters for troops to sleep. There was a post exchange [store] in there, a mess hall in there, a movie theater in there, you know. And a cleaning shop, everything was in this Marine Barracks. Up in the top floor was what we called the "Slop Chute." The Enlisted Club, you know. I remember one of our lieutenants made captain and was so happy he was shitting bricks. You know, 'cause he was a lieutenant for six years. They just didn't get promoted back then. So, he went up and says "Everybody come up to the Slop Chute, beers on me." He put $20 on the bar. Of course, beer was a nickel a bottle, so that's a lot of beers. So, we went up there and drank to him, and everybody toasted him and he said, "Now my next goal is to make Major" and somebody said, "Yeah, and that's about twenty years away!"

But, I was over finishing up in '63, I was in Vietnam and came back through Okinawa back to my unit up there and I remember talking to my gunnery sergeant, and I said, "Gunny, when am I going to make Lance Corporal?" which was E-3, and he said, "Well Hunter, it depends on the weather," and I said, "What do you mean?" He said, "When hell freezes over." Back in the States, I was already in two years, still an E-2 Private First Class. Promotions were just slow, and the average back in those days was to, if you got out after doing four years, you were an E-3 Lance Corporal.

Once you went through Parris Island, what was the first unit you were assigned to right out of boot camp?

I got sent to infantry training. That's a month long where they train you on all the basics of infantry, you know. Patrolling, building your fighting holes, compass work, and shooting all kinds of weapons. We got to shoot everything under the stars including the recoilless rifles, machine guns of all sorts. The Browning light machine gun, they still had .50 calibers. All kinds of weapons we got to shoot, which is a fun thing. In boot camp, we had the M-14 rifle. When we got to infantry training regiment, they issued us all the M-1, and it was a good rifle and everything, but it had more parts that were more fragile than the M-14. The follower on mine broke. I went and told the gunnery and he said, "Well we'll fix that." Every time we'd go through shooting, when you go from post to post shooting, shooting around windows and things, my buddy and my bunk mate would shoot his rifle, then he had to take his rifle apart, give me his follower, I had to take mine apart put the follower in there and shoot my rifle and then take it apart and give it back to him. But it taught me one thing—I could take a damn rifle apart! But that's what the gunny did to us to make sure we learned something, and, we got to sit down and take that thing apart and in 30 seconds have it back together.

What happened after you went through your infantry training?

Then I got out of there and then they sent me to Marine Barracks to start as a radio man, and learn, for starters, they taught us Morse code. Because we were supposed to be, that was still a big thing in the military then. I went through ten weeks of that, I think. We had a bunch of Marines in the class, and a bunch more Sailors because it was a Navy school. The Marines as you graduated, the guys who were really good at sending and receiving they'd put in the back rows, and the guys who were worse they'd put in the front row so the instructors could keep an eye on 'em, to see what they were doing. We found out pretty quickly in the back row back there, you could unscrew one of the screws, and put it over the two contacts, and you could short out everybody's receivers so they couldn't hear a thing! You'd hear "stop doing that!" all the time. Then, when we were doing our test you know, you'd sit there with the earphones on and a typewriter and typing in what you hear on the Morse code. Every once in a while, you'd see someone say, "Cut it out!" This was in Norfolk Marine Barracks up there. It was right down the street from where the carriers were all tied up.

Was it a good relationship between the Navy and Marine Corps?

That's where I learned how to play Pinochle. Once I learned how to play it and got overseas, I found out it was big in the Marine Corps too—used to have a blast playing it. So, I got promoted to Lance Corporal, finally, and went overseas to the Air Wing. I was a radio operator, so I just followed my lieutenant around back then. That was basically what I did. I handed him the handset when he wanted to talk to somebody. They used to have something in the Air Wing called a Pathfinder platoon. Every group had one, and they would bring the Pathfinder platoon in, and you would secure the landing zone ahead of the troop movement to provide security around the zone for whatever contingency might arise there. So, I did that a lot over there.

What Air Wing were you referring to earlier?

That was the 1st Marine Air Wing back in those days over there. First Marine Air Wing nowadays is in Camp Pendleton [San Diego, California], but back in those days, it was in Okinawa. It was my first time overseas—it was good, I enjoyed it. Beer was a nickel a bottle in the E-club. So, it was really good.

Did you enjoy the Japanese culture?

We didn't get much of the Japanese culture. We stayed on base most of the time, went to the bars. We had great organized athletics you know, baseball, softball, football. It was a lot of fun.

We had a contingent down in Vietnam at the time, in Da Nang, and it came down they needed another radio operator down there and they said, "Hunter, you wanna go?" I said, "Sure." They said, "you may get shot at, you know?" and I said, "Well, that's what the Marine Corps is supposed to be about?" So, I went on down in early December of '63 and stayed down there until it was time to rotate back to the States. Came out in late May and went to the States, back to an artillery battalion.

So, you were there from December '63 to May '64.

Yeah, so about six-months worth. Then I came back to the States, went to an artillery battalion and was like a section chief there, you know. Got promoted to corporal right away. Promotions were starting to pick up again. The Marine Corps was expanding a little bit in size, you know. So, I stayed there, went on a couple cruises to the Med [Mediterranean] and Caribbean. Had a blast. In fact, the last big Marine peacetime exercise ever. When we launched out of here to go out to Spain for the exercise, the entire east coast Marine Corps went. The entire Marine Corps Air Wing up and down the coast. All the Marine artillery units, the entire Marine division, entire Marine force troops. We took 28,000 Marines on the exercise. I looked out there, and my convoy had like eighty ships in it. In those days, you still had the big gun deck on the cruisers. So, we went over to Spain, had a big exercise. Killed 32 people in the exercise; friendly deaths, run over by amtracks. You know in those days we had UH-34 [Choctaw] helicopters. Those were our main lift helicopter. The artillery that could be lifted by these helicopters was towed mortars, they called 'em howitzers. They were light enough the helicopters could actually airlift them. They had a range of about 6,000 meters on them and shot 107mm shells. So, that was our artillery. Of course, each battalion had a battery of those things. The rest of the battalion was 105mm howitzers.

Do you remember where in Spain the exercise was?

It was in a place called Huelva. After the operation was over, they split up the units and let 'em, the Navy ships took different ways back, and we went to two places for liberty. We went to Porto, Portugal, where Port wine comes from of all things. We got off there, it's a funny story, and they said, "Alright who wants to go on tours here?" and a bunch of us raised our hands. We piled in a big huge tour bus and drove about three hours inland to someplace. They said we are going to go to Fátima, Portugal, where this miracle took place. So, we were driving and driving—I didn't know what Fatima was—and we finally got off and she said, "Here we are." We had a female tour guide. She said, "Here we are where the miracle of Fatima took place." We got off and someone yelled, "God, it's nothing but a church!" They were all pissing and moaning, and then someone yelled, "There's a bar over there!" We went to the bar and drank for two and a half hours until the tour was up. We got on the bus and went back to where our ships were. We made the bus pull over two or three times, guys were pissing against it on the side of the road.

You were talking about going over to Vietnam. Da Nang, what was that like? Your first tour in Vietnam?

It was interesting as hell. I was intent on seeing everything I could. In those days, still, we could have liberty to actually go out in town. But you couldn't take your weapons with you, of all things, and you could not wear a uniform. We wore washed khakis and utilities. I guess you call them battle dress or fatigues? So, we would walk out of the gate and there was this little Vietnamese village across the street. We called it "dog patch" and you could go over there and rent a bicycle for next to nothing. We'd all take a couple of cartons of Salem's [cigarettes] with us, and go downtown Da Nang, and you could trade these Salem's for a lot of Vietnamese money. They were desperate for the Salem cigarettes. We could go to a restaurant, and I could get a huge steak, you know, and for dessert I always got this huge omelet coated with brown sugar and high test rum poured all over it. They would set that rum on fire and the sugar would make a crust on top of it. One of the most delicious things you've ever had in your life. So, we would go over and do that, and we would go to the beach a lot too. The top non-commissioned officers would sometimes load up trucks and bring out tons of steaks and wood to make charcoal/wood fires, and grill our steaks right on the beach there and have a good time.

There were several exercises over there, operations actually, where they took us one time, and we flew down in helicopters and set up a perimeter. Well, later on that day, a whole bunch of every helicopter the Marine Corps had over there, about twenty-something of them, showed up, and we started loading up a bunch of [South] Vietnamese soldiers and Vietnamese rangers, and they all took off in convoys and went over about four or five miles over a hill, and landed over there—some shooting and fighting started, you know. Then they were coming back through, getting rid of all the wounded. The helicopters were shot to ribbons. By the time the day was over, we didn't have any more Marine Corps helicopters that could fly. So, I said to myself and my buddies, "What are they going to do now?" The next day, the whole sky was swarmed with Army Hueys that showed up. You know, picking up the Vietnamese rangers and soldiers, flying them back and forth to battle. Those Vietnamese in the back of the helicopters, some of them were wounded pretty bad... scared shitless and everything else like that. Eventually, they brought us back to San Diego and offloaded us onto the Grinder (the parade field) that taught the recruits how to march. That was how we got back.

Back in Vietnam during your first tour, was your base ever shot at?

Yeah, we got shot at a few times. We didn't see who it was, just had a good time firing back. Crank off a few. The gunnies [Gunnery Sergeants] were always running around saying, "What the hell are you doing!?" He was always mad at us. It was pretty mundane. It was a fun tour; I enjoyed the hell out of it. Of course, you enjoy every tour you don't get shot at. But, I went back to the States, to the artillery battalion, had a good time there, made corporal like I said, got out of the artillery battalion, went back to Okinawa again in '65. As soon as I got back in '65, they were starting to shuttle down to Vietnam, and they were going to move the 4th Marines down there to Chu Lai, back in '65, so I got assigned to the 4th Marine Regiment as one of their radio operators for naval gunfire spotter, you know. We landed at Chu Lai. Funny story is, we were loaded down so much, we couldn't climb, we had these nets on the sides of the ship, to climb down to the landing craft like in WWII. We were so heavy, we couldn't get over the side of the ship. So, they cut these booms and dropped the nets onto the deck, and we'd wrap our arms around the nets, on either side of it, they'd lift us all off the ship and swing us around and put us down into the landing craft like that. Because we had, you know, we had our field marching pack, I had a PRC-47 radio on my back. This thing is huge! It had two spare alkaline batteries in it that are this big [opened his arms about two feet wide] and each weighed about 14 pounds. Then I had my M-14, my helmet, I had a pistol too because I was a radio operator, my .45. I had six magazines with twenty rounds a piece for the M-14, and we all had a couple grenades too, they made sure of that. Then we had our big huge Mae West [personal flotation device] thing, wrapped around everything, and it was hot as shit too—sweltering.

We all charged to shore. We ran about one hundred yards up the beach through the sand and stuff. We were exhausted and looked as though someone had taken a water hose and just turned it on. Sweat just pouring off us, you know, because it was so hot and humid. We finally got there and started setting up camp and digging out our little fighting holes around the area. The funny thing was, the first three or four guys that got shot there were all shot by their own people, friendly fire. One night, we were up there in our fighting holes, and you had these little bitty "sound phones" that looked like this [made a gesture like he was listening to a cone], and I heard this rifle go "BAM!" "BAM!" and then, "Halt, whose there?!" He shot first and then said, "Halt!" We had several Marines getting shot like that, out running around taking a "whiz" or a "dump" or something. Finally, the battalion commander put out the word. "Alright, by God, nobody gets out of your hole at night. Once that sun goes down, nobody moves. If you gotta take a shit, take it right where you are. You gotta take a piss? Take it right where you are. Nobody moves until daylight." So, that put a stop to the friendly fire.

When was that?

It was May '65, Chu Lai. May of '65. We stayed there for a while. We had some good activity going on there. Three weeks after we landed, the Seabees [Naval Construction units] came ashore, you know. They had this expeditionary runway built already. A-4s [Skyhawk] started landing on it. But, you would see them when the fighting was going off all around the perimeter, they would take off and they would have jet-assisted takeoff [JATO] assist, in their wings, and they would arc back around, drop the JATO tubes into the ocean right beside us. Then they'd come back around and dive and drop their bombs about a mile and a half away. It looked like video from Peleliu in WWII, you know, when the Marines landed there, planes dropping bombs at the end of the strip. It was quite a sight. The artillery at night, you could read a newspaper from four miles away by the illumination flares going off. Artillery was just hammering all the time around there at night.

Did you guys leave your foxholes and go out on patrol?

Yeah, we set up patrols and went and established little base camps all around the area. Then they started doing the sweeps around the area, to start trying to clear 'em out. They eventually cleared 'em out. This was in July, and then finally in August of that year began Operation STARLIGHT, when they essentially cleared out all the Viet Cong there in that operation. Then it got real quiet.

That's when your buddy Lynwood Murphy was there too? That's when you guys saw each other?

Yeah. I stayed in Da Nang for the rest of that tour, until, around the summer of next year, '66, and I went back to the States. I was still a corporal, naturally! I finally got back to the States and, I swear to God I was bored to death. They put me in an Engineer battalion. Why that was, I don't know. I was just sitting there and my eyes were rolling around. They weren't doing anything! They weren't doing anything except sitting around, you know. I said, "God, I can't stand this." So, I went down and saw the guy that was in charge of supposedly recruiting and duty stations and assignments, and I said, "I need to get out of here and go back to Vietnam." He said, "Alright, they can always use a Naval gunfire spotter!" So I was there in Camp Lejeune for about three weeks, and then I was on my way back to Vietnam. I'd already been in Vietnam in '63-'64, then I went again in '65-'66, and then I'm going back in '66. I went back for my third tour in '66, and this time, as soon as I got ashore in Da Nang, we went over the ship again, and they offloaded us in Da Nang in the river there, right in front of the old Grand Hotel De Tourane. Bussed us over to the reception area where tents were set up.

As my luck would have it, some captain was really doing his job you know, really looking through his record books, he called me up there and I hadn't even been assigned to my unit yet, and he said, "I've been looking through these records here and you made the cutoff. You were supposed to be promoted to sergeant already." So, he brought me up there and some colonel came out and said, "You are a sergeant." I put on my sergeant chevrons and then the next day, I went back up again and he said, "By the way, we are sending you to Saigon, you're going to Air Naval Gunfire Liaison Company [ANGLICO]… their headquarters was in Saigon. So, I went down to Saigon, and checked in, and the ANGLICO unit was scattered all up and down the coast because they provided FAC services and Naval gunfire spotting services for Army and Vietnamese and Korean units too, that were there at the time. I did a lot of Naval gunfire spotting for the Korean White Horse Division [9th Infantry]. So, they flew me down to Saigon on a C-47 [Skytrain], and D.C.-3s they still had around. I went down there and the commanding officer (CO), Lt Col aviator pilot, because that's who the CO of ANGLICO of those days was a pilot, and he said, "Well, we've got several slots opening. Do you want to be a team leader?" I said, "Sure!" so they gave me Spot Team 35, which is up in Nha Trang, which is about halfway up the coast. I went up there and they had an Army unit there, lots of Army units were around that were responsible for our admin stuff like billeting and food and supplies and all that stuff, and fixing our radios. The Army had that setup, so, I went up there and, what we did, to tell you how adlib this whole thing was, I got my PRC-47 radio setup in this room they gave me in a little area of the base there, and I had an high-frequency [HF] link to my headquarters in Saigon, and had an HF link to the Navy off the coast in the 7th Fleet. As I'd walk around and liaise there with the Army advisors, and the Army units there, they would say, "We've got this operation going on here, going up and do this one here. We're going to need naval gunfire support for this." I would get on my HF radio and call the Seventh Fleet out in the sea and talk to them directly and say, "We need you up here at this time." They all had the call signs and frequencies, and so then I'd call Saigon and tell them what we were doing, and then what I'd do is go down there and go over to the Army, we had an Army [Thomas-Morse] O-19 biplane company there, and an Air Force detachment for O-1s [Bird Dogs], for their O-1 unit. So, I'd go talk to them and see which one of them was available to go fly. I'd get in the O-1 and take off and go fly up there and contact the ground unit and talk to the ships, and start shooting naval gunfire. They would shoot 'em out in one day.

It used to piss them [the Viet Cong] off because we had some rocket ships, that could hold like three or four thousand rockets that had a range of ten thousand meters and they could get right up to the coast, on the shore with those flat bottom ships. We'd start shooting naval gunfire missions and, we'd just keep going, and shooting. One time we shot I think five-hundred rounds at these twenty guys running around. They were quick little buggers. I must have shot, some days, twenty or thirty gunfire missions in one day, and over a period of time, hundreds to thousands of those things. But we were shuttling back and forth in our A-19s or the O-1 Bird Dogs with the Air Force pilots flying them. They were some damn good pilots. They knew their business. The O-1s, with a crosswind, would come in sometimes landing sideways almost, and at the last second, he would snap the rudder around and put the wheels down, wing down and so forth. Those guys used to scare the crap out of me, though. When I first started flying, we had the windows underneath the wings, so you could look out and if we were on the coast, working with a destroyer trying to get in close or a rocket ship, would get in close, or even the Coast Guard showed up! They had 40mm cannons on those ships! They would carry a bucketful of smoke grenades in the back of the plane. We'd get real close and throw out a smoke grenade to mark the target. It was great fun doing that, but those guys the first time I ever flew though, the first day, he said, "I want to tell you something. You know what popcorn sounds like when it's popping, right? Well if you ever hear popcorn going off, we're getting shot at because that's what it sounds like." Popcorn popping in a popcorn machine.

How was your experience different during the second tour in Vietnam compared with the first?

It was so different from the first because now there was actual fighting going on. The first time, you weren't even allowed to go out in town in a uniform, I guess we weren't supposed to be there. But the second time was interesting, there was some action, but in August of '65, when they cleaned up that Viet Cong regiment, things got real quiet, and from then on they were just conducting sweeps just to keep things going smooth. There just wasn't any resistance from them in that area, because it wiped out the Viet Cong in that area.

Is naval gunfire spotting the same as being an aerial observer?

An aerial observer does that as part of their job. But, they also do artillery from the air, controlled airstrikes from the air, and naval gunfire. But, if you're a ground Naval gunfire spotter, pretty much all you're talking to is naval gunfire.

What training was required to do that?

I learned it on the job. The Marine Corps had a school for that, but I was already in Vietnam learning it. As a matter of fact, because I was doing it, we were doing reconnaissance flights too in the O-1 Bird Dogs, or the O-19s, after we did it for over one-hundred missions, you get one hundred and fifty flight hours of combat, you got awarded your aerial observer wings and designated as an aerial observer as a secondary Military Occupational Specialty (MOS). So, that's how I got to be into aviation to start with, but I was still a grunt MOS, so to speak. So, like I said, I did the rest of that tour in ANGLICO over there, doing all that kind of stuff, and did a lot of ground operations too with the Korean Army. Like I said, the White Horse Division and also the Tiger Division [Capital Mechanized Infantry Division], were in the southern part of Vietnam, so I got assigned to them a couple times for exercises and for operations.

One big one we had one time up around Quinn Yan area, we loaded up a couple battalions of Koreans and loaded them in UH-1 [Iroquois] Hueys, by the way, and chalk full of those Koreans. They jammed them in! We flew out to the mission and I had myself and one other guy with me, and I had a PRC-25 radio, which was such a big advance over what we used to have. We had a day's worth of combat rations [C-RATS] and my poncho and some socks and my shaving gear. That's all I carried, with my weapon of course, but you had a ton of Koreans around me and they had weapons on them, those guys would kill everything in sight. We flew into the landing zone, and it was a hilltop, and they [the pilots] didn't want to stop, so they dropped down to about a foot off the ground and coasted along at about two to three miles per hour and everybody just bailed out. Huey after Huey came in doing that, and some of them took too long and their knapsacks were falling down the hill a few hundred yards, so they were scrambling down the hills and hauling them back up, and then it started pouring down raining. Then night came, so I just sat myself back up against some thick brush, put the poncho over me, you know, and just sat there all night in that stuff. I had my radio on, had it jammed up underneath my head so I could hear it, to see if anyone was calling. But we went to several little ridges that day and sat up on a promontory looking out, and we saw some of the Viet Cong running around down there and everyone started jumping there and yelling and shooting at them like a bat out of hell. Hosing 'em with machine guns, you know.

The Korean commander I was with, he was in this little pup tent, like our two-man tent, and he had three or four radio operators standing outside, and he would yell at them, and they'd hand them a handset through the tent, then he'd yap at it for a while, and hand it back, and he would yell at them again and they'd hand him another handset, and he'd talk for a while. One time, he yelled for a handset, and a guy didn't respond, so the Korean captain crawled out of that tent went over there and grabbed that guy's helmet off his head and knocked him out right there on the spot. Just knocked him flat out!

Were you the only American?

Yeah, at that time I was. I started with another guy but he went back to the base camp where the regiment was. I was there by myself.

I bet that was pretty interesting working with the Koreans.

Yeah, the overall controller for us Marines was a captain named Claire Everoll, was with the Vietnamese regiment you know. He was really our liaison officer. Me and two other sergeants were out there and we were the Naval gunfire spotters. They were supposed to be Marine officers, but we didn't have enough officers back in those days, so he filled in. All the NCOs that we had that filled in, they did a really good magnificent job, and I kept telling all the officers when I get to talking to them, I say, "Your NCOs can do more than you think they can. Just give them the authority and give them the leeway to go do it. They can do good jobs for you."

How did you go from being a communications/radio operator to naval gunfire support?

Just happened to fall into it. There was no rhyme or reason to it. Just when I made sergeant, they sent me to ANGLICO, like I said, and the CO said, "Well, I see your record, you look pretty good. How about running your own team?" I said "You got it Sir; I'll take Spot Team 35." He flew me up the next day to Nha Trang, and he said we got a sergeant there right now, he's going, you take over. I went over there, got in the Jeep, and visited everyone I could and said, "I'm the new guy on the block here." We did that for several months until that Navy lieutenant showed up, and he took over as the liaison officer. This whole tour lasted from '66, and I left this tour in Vietnam in '67. We had a good time, did a lot of naval gunfire shooting. I saw an article in the Marine Corps Gazette, and it talked about the structure of ANGLICO in Vietnam. They showed the Spot Team at Nha Trang, and I wrote them a letter and said it wasn't "Us" and then "somebody else," it was just "Us," our five-man team. We went to all these cities and did all this. We had no other teams there. We covered the whole two hundred and fifty miles of coastline.

Can you explain ANGLICO?

What they do is—it's a pretty good-sized unit, they farm out teams to support US and allied forces, such as a joint operation with the Army. They'll send a Naval gunfire spot team to an Army unit to control the ships that shoot naval gunfire. Or if you're Naval air, you help coordinate Navy air. So, we did it all, and we did it for the Vietnamese army over there, the Koreans who were over there at the time. We did all their air and Naval gunfire spotting for them. ANGLICO worked directly for Fleet Marine Force, Pacific.

What was your third tour in Vietnam like?

It was good, it taught me a lot. I learned how to really operate because I was on my own a lot. You learn how to just go find, if you want to do something, just go find some way to do it. I had to have comms with the fleet, to get ships and talk to them, about when to be, where to be, and how long we're going to use them and so forth. They would tell me when they were going to be there and how much ammunition they had on board. So, I went and just figured it out. I went down to the Army and finagled them out of giving me two huge truck 12 volt batteries and took them to my room up there in the Army compound and I got some wires and wired them up in series, so then I could have me a charger cable, I plugged that into the wall, put the cables in the bag so I had constant power forever in the room I was in. Then I took the power line and ran it over to my HF radio and so I could leave the radio on constantly, twenty-four hours a day. Then I took an antenna and ran it out of the window up to the top of the building, put a pole up and put the antenna up there and then I had comms with Saigon and the fleet, you know. All out of my room. Then we took the maps, covered the walls with maps, you know, 1:50 maps the whole area. Had everything laid out there, so we could keep track of who was doing what and where the units were out there. That's how we sort of kept track of it. We'd put little marks on there, little notes, about "We did this here and this is what we saw" and so forth, and sent the reports in monthly into Saigon-here's what we did down there.

The lieutenant you were speaking of, he was down in Saigon?

He was there, but then he came up and joined us. We kept asking for lieutenants, and they finally sent us one. His tour was up off the *Intrepid* [USS *Intrepid* (CV-11)], the aircraft carrier, where he was a gunnery officer, and he wanted to come do something adventurous so they sent him zero in that tour. I said, "Guys, we're going to be distinctive, we're with the Army now." So, I went out and got us all scarlet berets. I had our call sign written in script into the side of the berets, Scrappy Two Six Charlie was our call sign. And then, I said, "We got to be better than this." So everyone got different weapons. So, one guy had a shotgun; one guy had an M-14; one guy had a Thompson submachine gun. I got myself at that time an M-2 Carbine. It could go automatic as well as semi-automatic. Then I went down to the armory and got myself a .357-magnum revolver, six-shot revolver, with a six-inch barrel on it. I went and bought myself another one, and so I took my gun belt and modified it so I had dual guns like a western outlaw. That's how I walked around the combat zone. Two six-shooter, .357 magnums! We were like swashbucklers down there in that unit, because there wasn't anybody from the Marine Corps there to fuck with us. I think the CO flew up about two times in his own plane, got some kind of twin-engine plane, to come up and visit.

We were always looking for ships to shoot naval gunfire with. Then we found out down in Cam Rahn Bay, which was south of us. There was a seaplane tender down there. I just happened to run across some messages. Anyway, they were getting ready to send it back to the States. They were getting ready to decommission it, you know. Well, I looked up in the Janes book and found out they had a five-inch cannon on that ship, four of them. I got ahold of Charlie Thompson and said, "Let's ride down to Cam Rahn Bay and see if we can talk to them and see if we can get them to shoot a mission on their way out of Vietnam." So, we jumped in our Jeep, me and him and a couple of guys, and we were heading down the road to the bay, which is about fifty or sixty miles away. We're driving along and following this Korean truck in front of us from the Tiger Division I think it was. There was this little girl there, like eight or nine years old, like she's getting ready to cross the road, and he swerves over and knocks her over with that truck. Charlie said, "That son of a bitch, let's get him." So, we started chasing after him in our Jeep, blowing our horn. He wouldn't slow down. So, we started leaning out of the Jeep and shooting at the truck with our guns, trying to shoot the wheels out! I guess we scared him because he stopped. We ran beside the truck, and he was trying to grab his rifle, and our Lance Corporal took his gun and knocked him out with it, and we threw him in the back of the truck. So, we turned around in that truck, and one of our guys drove. We headed back to where he hit that girl and as we went along, we saw this other jeep pull up, and it had Koreans in it and had a couple officers. We pulled up and stopped and asked if they spoke any English, and they did. So, we told them the whole story, he got on the radio and started yacking with that Korean truck driver and he started beating the fuck out of him right on the spot. Just hammering him right in front of us. He goes, "Alright we found out she's over here in the hospital, she's okay." So, we said, "Alright, see you later." Turned around and headed back down to Cam Rahn Bay. We went and got aboard the ship, went to see the ship's captain, put our proposal to him. He started laughing. He said, "These guns haven't been shot in 20 years. We don't have anyone on this ship that even knows how to load them." We drove back to Da Nang, all happy as hell we got to chase that truck and got to shoot at it. We emptied all our guns out doing that!

So, after your experience on that deployment, what did you do next? Did you come home? Did you get a chance to have some leave? Seems like you had been running pretty hard.

I was having the time of my life. I went back to Georgia and visited a couple times. My dad said, "Where you been?" By that time, I had extended my enlistment for a year to go back to Vietnam [the third time], and so my enlistment was coming to a close. So, I had to re-enlist. I already had five years in the Marine Corps when I came back in '67. It was a four-year enlistment plus a one-year extension. So, I went down to Saigon and talked to them, and they said, "Well what do you want to do?" and I said, "I'd like to go to a language school." They asked which three I wanted, so I said, "Well, I'd like to learn Russian, or Serbo-Croatian, and one other I forgot." They re-enlisted me on the promise I would go to language school, but only if they are available. So, I left Vietnam and went home, and then I went to Company B Headquarters at Henderson Hall [Arlington, VA], because that was who controlled us administratively. I checked in and got assigned to the school and went and talked to the school and everybody and they said, "Well, some of these classes are unavailable." They told me I could take North Vietnamese! So, I went and took North Vietnamese for a year there. At eight o'clock every morning to 1450 every day, I'd go to class and learn North Vietnamese with different North Vietnamese women, you know. Because

North Vietnamese is different from South Vietnamese. Much different than southern and northern accents here in the States. Anyway, the guys in the class, had a couple officers in the class with us, we'd go down to the bar and have a couple of beers at lunch, play some Pinochle, then every day for class, we had to learn two dialogues. So, you'd learn a dialogue and I'd learn a dialogue, and we'd get in front of the teacher and say them back and forth and swap off. Then, every evening we'd have about forty or fifty new vocabulary words to learn in the language. Which turned out to be, I found out that Vietnamese is one of the easiest languages in the world to learn, because it's pretty straightforward. So, I stayed there for a long time, just having a good time in [Washington] D.C. Had several lieutenants in there. Strangely enough, one of the corporals in there, kept telling the lieutenant he wanted to take his Corvette and go to Canada with it. He kept laughing at it; about four days before we were done with the class, he took the Corvette up to Canada, that corporal did! Never heard what happened to him after that—he just disappeared. I got myself an apartment really close to there—it was in Fairfax, VA—there were three lieutenants that lived above me, doing some school. I used to go out in the evening and they'd be above me drinking. They used to talk to me about experiences and I'd tell them, "You're going to Vietnam. They are going to kill your ass! You got your wills ready? You aren't coming back lieutenant!"

After school, I went back to Vietnam and checked into 3d Marine Division. Because I had Vietnamese language for a year, they sent me up north with a Vietnamese battalion. I did airstrikes and naval gunfire for them. We still had, at that time, big gun cruisers at the time, the *St Paul* [USS *Saint Paul* (CA-73)] and the *Chicago* [USS *Chicago* (CG-11)]. They shot them right up the coast for naval gunfire. By that time, the command and control of that war got so screwed up it was unbelievable. I was up with my unit, the Vietnamese battalion, and when I called in naval gunfire, I had to call it in, had to go back to the Fire Direction Center and then what they would do is route it over to the Vietnamese regiment which was headquarters there, and they would call up to the battalion commander that was standing next to me, and he would say yes you can shoot, and it would route all the way back to me. That's how long it took to get a mission shot in those days.

How long did that take?

About 25-30 minutes. That's not even dropping ordnance, that's just approval. Then I'd call the ship and say, "We're approved," and they'd start shooting their eight-inch guns toward us. It was on the border of North Vietnam; the camp was called Alpha Two. We had a Vietnamese battalion dug in there. We'd set out ambush patrols every night from that place. We also had an Australian with us for a while, too. They would have a code word, for example, "Candy." So, if you called in and said, "I'm at Snickers, up two right two," then they would know you were using a certain checkpoint for map position. They would change those about every two or three days. Or another example, "Car," they'd say, "I'm at Pontiac, left two down two." For grid coordinates.

What year was this?

I went to Vietnamese language school, started in '67, so I got out in October of '68, flew back over in October '68 and checked in to 3d Marine Division. It was a year deployment; I left November of '69.

What was that experience like, your fourth tour?

It was great; I learned a lot again. We had to do road sweeps every day. The Army was up there at the time, and they had a tracked vehicle called a duster. It's got a little turret tub at the top with twin 40mm guns in it. And that's what they used for protection on this road tour, to clear the roads. It was during this tour, toward the end of it, because I was promoted to staff sergeant in April of '69. So, I had an epiphany like Paul had on the road to Damascus. I was out there doing ground FAC work, and I was north of Vandegrift Combat Base, on a hill there, dirty as hell. They farmed us out to all the various units there. I was up there and this CH-53 [Sea Stallion] had been circling around us and dropped off a sling load of C-RATS, and five-gallon water cans. It was just dusting the air, got in the back of your neck. That CH-53 dropped and flew away back to the east and one of the guys said, "You know, that SOB, he's gonna sleep in a cot tonight, and he's going to drink a cold beer." And this light shined from the sky, and I said, "I gotta get out of this line of work and do something like that! These guys are trying to kill my ass all the time, I gotta go do something else." Talking to my good friend, he said, "We got a new Marine Corps bulletin out looking for people to be warrant officer air observers." He said I should put in for it since I already had my wings. So, I went ahead, and he helped me fill out the paperwork. Just as I was getting ready to transfer back to the States, they started pulling the 3d Division out then, late of '69. We were burning piles of material, piled with grenades, setting them on fire. Just mountains of stuff that accumulated during the years of Vietnam. Piles of old radios and stuff. They didn't turn 'em in, they just burned them. It was a waste.

So, I got my orders to head back to the States, and to be an aerial observer, you had to be a warrant officer, and to be a warrant officer in those days, you had to go to officer candidate school (OCS). So, they sent all the warrant officers in those days to OCS, so I went there. By early December I was in OCS. I checked in, and I thought man I'm an old salt and check in and tell them, "I'll be back tomorrow. I'm going to go out in town for a while." Next thing you know, I said, "Holy crap, it's nothing but boot camp all over again." Waking up in the morning and holding your sheets out. "COUNT OFF!" I said, "Dang I didn't count on this stuff!" OCS for us was about six weeks, I think. After a while, they had a lieutenant for a platoon commander and a sergeant instructor. After a while, they finally quit screwing with us and said, "We're going to get through this and get you out of here."

Was it all warrant officer applicants in your class or were there officers as well?

Just warrant officers; about three platoons of us, 300 or so. We actually had two or three guys flunk out of it. They just were not physically fit! I couldn't believe they got in there. You've gotta be able to run. You don't have to be a speed demon, but you've got to run! And you've got to do the obstacle course. They couldn't do it! I thought, "How the hell did you get where you are?!" But we got through that. Finally, graduation day came. I got promoted, and was an honor graduate. Afterwards, I went back to Camp Lejeune, checked in, and to tell you how things were in those days, I checked in and the adjutant brought me to see the regimental executive officer (XO), and he said come on in Gunner, which meant a line officer in the Marine Corps, and we started chit-chatting, and he said, "Do you want a welcome aboard top?" He pulled out a bottle of scotch and poured us both a shot. He saw my aerial observer wings and said, "I know about you air observers. You sneak over to the Air Station and hang out with all those flyboys all the time." He said, "We can't have that," and sent me to artillery school at Fort Sill, Oklahoma! I was there for a while, and did a couple training revolutions with the unit—a couple of operations. Learned a lot. I was there for a total of about three and a half years. I eventually bounced between VMO units. Helped to develop a lot of the aerial observer training programs, and ended up in VMO-2 leading up to the beginning of the Gulf War.

Following his tours in Vietnam, Guy Hunter was assigned to VMO units on both the east and west coasts for nearly two decades. Due to his extensive experience, dating back to the Marine Corps' earliest implementation of aerial observation, he was responsible for developing much of the aerial observer training programs. He went on to train several classes of aerial observers for several years, eventually ending up stationed at Camp Pendleton with VMO-2. Handpicked to deploy leading up to the Gulf War, he was one of the first to arrive in Saudi Arabia among his OV-10 compatriots. The men in his unit endearingly referred to him as "Great White," both an homage to his experience and a tribute to his full head of white hair.

DESERT STORM

What was your impression of VMO-2 when you arrived?

Well, I knew most of them, I trained half of them. I flew with them over one place or the other. They were all good guys; I liked them, a lot! In fact, when we got back from DESERT STORM, I told them all, I said, "This is the best set of captains I have ever been associated with in a Marine unit." They have a bunch of captains and they were all eager and proficient and efficient in every one of their fields, they knew what to do!

Talking about the advance team to the Persian Gulf, how did you get notified?

I was out jogging around the airfield at Camp Pendleton, and I got back in, and they said, "The CO is looking for you." He said, "We are all going over there, but I want you over there, right away, with the advanced party. You will help and get everything set up." So, that is when I drove up and picked up Mary and said, "Goodbye, see you later!" Got on a C-130 and started to head east. The rest of the guys packed up and they flew the OV-10s all the way across the Atlantic, up the eastern seaboard, over through Baffin Island, Greenland, over to Scotland, then flew through Europe toward Saudi Arabia.

How long did that take? And what were the training and meetings like prior to the start of the war?

It was several weeks, like four or five weeks. It was over a month, but I was in the advance party so I wasn't part of that group. For training, I was Assistant Operations Officer for the group with a colonel in charge of the group. So, every Monday, I had to go down to the group, which was at this other Saudi Arabian base [Dhahran Airfield, King Abdul-Aziz AB] and have a big meeting with the American flyers, Saudi Arabian flyers, the French flyers, Air Force flyers, and then there were Army flyers. Then everyone would divvy up who was going to fly at this particular area and this particular time and this particular altitude. Everyone would get their specific air space. Then the next Monday, we would go back and confirm and plan for the following week. Get the training spaces set up. It was an interesting time. I was getting a lot of flight time in. There was a lot of dead reckoning, because all we had was TACAN (Tactical Air Navigation System) in the back of the OV-10s. We also had a new long-range navigation (LORAN) navigational system on there. This was something for sailboats, not OV-10s. There was lots of good training and shooting at the ranges.

What was interesting at those meetings was that they had a huge room and at the end it was chaired by this Saudi Arabian pilot colonel. I told the CO at the squadron that I was tired of doing this every Monday, I wanted someone to trade off with me, so he got this pilot. I forget his name but everyone called him the "man with two brains." He was called that because you ask him what time it was and he would tell you how to make a watch. So, I took him down with me to the meetings and during the meetings, once in the morning and once in the afternoon, they hired a bunch of Filipino workers there and the Filipino workers would come into the meeting room and put out these beautiful saucers and cups, really nicely designed gold and hand painted and they would pour this really good sweet tea in everyone's cup. It tasted good and refreshing too and then they would come back in the afternoon and do it again. So, we were there one morning, and this USMC guy, call sign "Rainman," because that is what he was like you know, Rainman. So, I said, "Rainman, these cups are beautiful! Rainman take one home with you. No one is looking, while everyone is talking, slip one of those saucers and cups in the side pocket of your flight suit." When no one was looking he slipped the cup in there you know. When we came back in the afternoon, I told Rainman to take another one when no one was looking and he got away with it. He had two complete cups and saucer sets. And then, when we pulled back into the compound after driving back that evening about an hour and twenty minute ride in the truck and the Colonel was standing there where we parked our vehicle and got out. And he said, "How'd things go?" And I said, "Sir, you got to do something with Rainman, he is gonna embarrass us and get us all in trouble, I took him to the meetings and he is stealing their cups and saucers." He is saying, "You made me do it, you made me do it!" Colonel said, "What's wrong with you Rainman?" He is as mad as hell! "He made me do it, Colonel, he made me do it!"

Did Rainman ever return the cups?

I don't know, I don't know whatever happened to Rainman, I talked with him once, he lives up North. He was flying for the airlines. He was an Aerial observer before he got out, he went and took these private classes to be an airline pilot.

You mentioned some of the training you were doing? Dead reckoning, what other type of training or missions were you guys doing?

Well, we did a lot of ordnance shooting and also, we did a lot of inter-training with other air platforms, you know FAC in FAC-A training, out in some of the areas, with [AV-8B] Harriers and F-18s [Hornet], some planes off the aircraft carriers would come in occasionally. We had Navy guys at that meeting for air range times. We had some pretty good training there. We got used to flying in the desert and not seeing a damn thing. We would fly around and there would be a guy down there with a big huge bed sheet walking around with a bunch of camels. And I used to say, "What the hell do they eat? I can't see anything down there for camels to eat! You got to eat something, you know." And what was amazing too is you can find these little dirt roads you know, and you just keep following it, and it gets bigger and all of a sudden, there would be this little town there. It looks like something out of the movie Beau Geste, with the French Foreign Legion, with the walls and everything. That is what it looked like. Just out in the middle of no damn where! It just looked like something mystical.

What was the atmosphere like around the base at this time? Was there a lot of uncertainty of what was going on?

No, they were pretty much looking forward to it; they were looking forward to it. And it came to me that November, it was the Marine Corps birthday, November 10th. And so, the CO, they had the squadron have a little birthday ball, the CO ask me if I would do the speech for them this year. I told them a good speech about tradition, honor, and all that kind of stuff and training is good for you and I got to the end, and I said, "Now, I want you all to understand something. The shooting is going to start pretty soon and I want you to be prepared. I never saw a Marine fail to do his duty and when the combat starts none of you will either." I said, "You don't have to worry about a thing." One guy said, "What's it like getting shot at?" I said, "Don't you worry, you will do just fine."

How did you hear about the war starting?

Oh, they told us in advance. The commanding general came by to visit on the 16th of January, and he said, "We are gonna start launching strikes tonight!" And it was the 17th there. So, I got up and went outside; you could see all the planes going north. And what brought it home for us was that I was the OPSO (Operations Officer) and I was writing the schedule for the squadron there, and I set up for me and Acree, my CO, as the pilot, to have the dawn launch. So, we got up at the break of dawn and headed out to look around. We had reports of artillery, Iraqi artillery firing from southern Kuwait to our forces. So, we headed that way and checked in. What brings it on home for you is when you cross the line, and you are doing your combat checklist, you know turning off all your running lights, turning on your super-secret transponder codes. You know that kind of stuff.

So, we went in there, and we were the first launch of the OV-10s on the 17th of January, so we tooled on up the coast of Kuwait, looking around at what we might see. And we could see the Iraqi artillery batteries in a perfect row. You could see the muzzle flashes where they were firing. So, the first thing I did was to get on the HF radio, I called the Direct Air Support Center and no response as usual out of those guys. And so, I got on the HF radio for our squadron at group ops and I said, "Run down to the group ops and tell them we have some Iraqi artillery batteries spotted and fixed, here is their location, tell them to start sending me some flights of Harriers and as soon as they get here. We could knock the snot out of them."

And I said we are on this frequency and I forget the code names. Soon after that, I heard flight check two, three, four is up. Their Harriers were at 12,000 feet, we were down at around 5,000 feet. I said, "We are in for the mark." We were flying a green OV-10 at the time, we didn't have the desert paint yet. We fired a Zuni rocket. The Harriers started rolling in carrying 2,000 pound bombs and rippled up and down the artillery battery and started to blow it up. They ran out of bombs and I said, "Ok, now come in with your guns." Another Iraqi group showed up and we blew the hell out of them too with the Harriers. We had a couple of really good missions there. Destroying two Iraqi batteries in an hour and a half.

In fact, it was the very first Harrier airstrike in combat, ever, it was that morning. So, we got to control the very first ones.

So, you took off multiple times then?

No, we went up early that morning and we had about three and half hours total on station time, and we landed again. We were real short on aircrew then, because VMO-2 had not arrived in full force. In fact, some of them were flying in that day, some were arriving. So, they came upstairs and I briefed them because I knew them all. I trained most of them and trained with them. And, I sat on the desk and started taking all the reports, went down and debriefed with the group, told them what went on and then when on to write the schedule for the next morning, because we were scheduling from midnight to midnight. So around midnight, I got off the desk and turned it over to someone else, and I went into my tent which was real close, got about four hours of sleep. Then me and Acree got up and took off again for a dawn launch.

CAPTIVITY

And that was the morning of the 18th? Let's talk a little bit about that morning when you woke up, how did you feel?

I was tired, and it was cold, because it gets cold there in December and January, you know, it really does. People think desert as hot, but desert doesn't mean hot, it means dry and cold too.

Do you remember what you had for breakfast?

Some garbage, they never had good chow there. Even the coffee, I would complain to the mess hall, I would say, "For God sakes!" [referring to the coffee] Mess Hall replied, "Well, we have to make it, so it's down there." I said, "Well, you don't have to make it six hours early!" The coffee was lukewarm at best, but you know, there never was hot coffee the whole time we were there.

So, you woke up and you and Acree got ready?

I took a quick shower, went up to the mess hall and met there. We already knew what flight number we were, so we just sat there at the little wood table they had outside, drank our cup of coffee, went upstairs, grabbed our stuff and headed out to the airplane. And that's when I had such a foggy mind that I forgot that my wallet was still in my pocket. I wasn't sanitized like I should have been.

What was the weather like that day, that morning?

It was a little overcast, you know, thin clouds. Chilly and it had been raining a little bit, so the ground was wet.

Was there anything going on that morning, any other flights that were taking off?

Yeah, we tried to keep an OV-10 up 24 hours and so what they would do was go up and monitor the border and make sure there was no heavy incursions coming our way you know. And they reported that there was some Iraqi rocket fire coming our way over the border and I got a rough estimate of right where they were, drawing a circle around our map. We headed north to see if we could try and locate them. We had them fixed, and we decided to radio it back, and they said, "We have some A-10s [Thunderbolt II] on the way," from the Air Force. I got that word from the AWACS guy. I guess that is who he is, I don't know. I think they call it STARs or something?

JSTARS [Joint Surveillance Target Attack Radar System]?

Yeah! And so, we were up there looking for that. And all of a sudden, they were waiting for a vulnerable point because that's when they started opening up and that's when we got shot by a missile.

Can you describe that for us if you recall from the moment you guys saw your target until...?

Yeah, we had the target in sight! We were just waiting on somebody [referring to the A-10s] and then all of a sudden, in a half instant, I saw that we were being fired upon. And just a half a second at the most and then all of a sudden that—well a lot of folks like the Intel guys, they thought that it was an SA-8 [Gecko, SAM] because it had separate optics that it could guide in, you know, from some type of wave runner so you don't get any radar warnings on your gear. So, that is what they probably think it was, an SA-8, but I am not sure.

I felt this huge "WHAAM!" on this side of the plane [body language indicated hit from the left side] and I was knocked out cold, you know, just laying there in lala land. And I came to and the plane was going downhill and we were on fire too. I said, "Holy Shit!" I grabbed the stick and tried to pull the stick back and it was stuck, nothing would work. And I said, "Well, it's time to GET OUT" to myself and I went to grab it [the ejection handle], and I lost consciousness, and I was out like a light again. And the next thing I know is I woke up in my shoot coming down at around 2,000 feet I guess and the first thing I thought as I gained consciousness was I gotta steer this thing and get to the Gulf and get my little raft out. That way I could paddle my way back to safety. That is what I was thinking you know. You know I was thinking they will never catch me if I get in the water. Because I can paddle like a bat out of hell! But after a little while, I realize that it was useless, I was coming down over dry land and it was flat and you could see the entrenchments, it was in the middle of an Iraqi division.

So, I knew I was in for it. That was one of the first things I said, "Man, you are in for it now!" And I landed there, took off all my gear and tried to call on that useless survival radio. That damn thing never works. I never had one that worked. Anyway, we landed, and I had a slight concussion, so I was seeing double for a long time. And I saw in the distance these two people just walking towards me. I started working away from them, looking back and finally I heard Colonel Acree say, "Hey Hunter!" It was him. So, we joined up and sometime before that I remember taking my wallet out and scooping out a depression and putting sand over it, left it there. We joined up and stood there for a few minutes wondering what the hell to do because we were both in a daze. He had a nasty lick on the side of his head. And he was bleeding pretty good and I was bleeding like a stuck pig. When I had been ejected out of the OV-10, something had cut my eyelid in half. I could close my eye and see out of it because the eyelid was sliced. So, after we were on the ground, a little truck drove up, like a little Japanese Toyota thing you know, little 4-speed. And five of them jumped out, not with AK-47s but AK-74s, went through the hands up routine.

You mention a concussion was there any other types of injuries on you or Acree?

He had a piece of shrapnel stuck in his head for a long time, and I think we finally got that out when he got back to Bethesda [Maryland], to the States. Yeah, so we joined up [with the Iraqis] and the first person they took us to—they took us to this underground bunker. I guess it was like a battalion headquarters. And their version of a corpsman came out and started to wrap a bandage around my head. And then these guys tied us up with some type of cloth thread with our hands behind our back. Then they put us back into the truck and then we drove away again for some distance, I can't remember or tell you how far now. We went to a division or corps headquarters because it was a huge underground place, a huge place. One of their officers came over and he said, "What base do you fly from?" and I said, "I don't think I am supposed to tell you that." And he said, "Ok" and that was the end of the interrogation. [laughter] It's funny you know!

He was rubbishing through my pockets and I had a stick of gum and he said, "What is this?" I said, "Chewing gum! You know, chewing gum!" He said, "Ok!" and he took it and started chewing on it. And then they took us to another place and I thought they were going to behead us there because I could hear paper being spread out over the floor, like butcher block paper. And then they made us kneel down on it and I can hear them talking in Arabic and I could hear what I swear to God sounded like swords being pulled out of the scabbards. I thought they were going to do a beheading; it was one of their things. And I was thinking, "Well, this is it!" They yelled and beat up on us for a while, then they tied us back up. They put handcuffs on us, and cinched them up real tight. It was cramped hard! We were in pain. We stayed there a few minutes and they loaded us into a truck again, some type of vehicle because we were blindfolded. And we were heading as it turns out, heading north to Basra, into Iraq.

You know what worried me most was that I had been down to the group that night, you know, before I went to bed and they were planning on this strike right where we were that morning. And all I could think of was, boy nothing attracts attention more than a moving target! Here would be a couple of Harriers rolling in and with 25mm cannons would come in on top of us. I was out sweating bullets more than anything else at the moment. You could hear bombing going off all around too! I mean it was heavy in there because our air units were laying a licking on them.

It was constant thunder, and jet sounds overhead; it was a busy day all around. We got into Basra. They took us into a big huge room somewhere downstairs. It felt like it was below ground. And as it turned out I could hear Jeff Tyson over there, an Air Force major at the time and he was being interrogated. He was talking to them, and he was just telling them bullshit stuff. Like where you from, "I'm from never never land. And Peter Pan is my CO." Shit like that, he was telling them. Not that specifically but similar crap like that. I was thinking, what the hell is wrong with this guy, they are going to kill his ass. And finally, the Iraqi said, "Listen, I don't care what you tell me. I will write it down and take it, but you better watch out because in Baghdad they won't believe this kind of stuff and they will do something to you." You could tell he didn't want to do anything to us. You know the time we come in contact with regular Iraqi Air Force officers or line officers, you could tell, you could really tell, they didn't want any part of this. I guess they felt captive to the situation. And so, he [Jeff Tyson] was telling them stuff for a while, and then they brought me in there and asked me a couple of questions, nothing too important and they put us all in holding cell throughout the night. We got up the next morning and got into some type of van and started trundling again.

Ah yeah, and that night, they brought in a couple of more guys. A couple of Italians, I think, and one of them was hurt pretty bad, bad leg. They were tornado [Panavia Tornado IDS] drivers and Jeff Tyson was an F-16 pilot. So, they loaded us up and took us the next day into Baghdad and to a huge underground complex, you could tell we were going down and down and down on this ramp. You could tell it was underground because you could hear these fans moving air around—mmmmmm—constant movement of air pressure. We stayed there for quite a few hours, and they just laid me out next to the hallway, next to someone who I couldn't tell who. You could hear these Iraqi's walking back and forth. They would haul off and just kick you in the head. It knocked me out a few times. Of course, we had the hell beat out of us coming up to Basra and Baghdad in the vehicle.

You could hear the missiles just blasting off outside; it sounded like SA-3 [GOA, SAM] maybe, big enough that is. Nice size ones. And then, the anti-aircraft batteries banging away. Their heavy anti-aircraft stuff and the missiles lifting off. You could hear explosions and the sounds of jets and bombs going off. Of course, you couldn't see anything, we were underground and blindfolded to boot. Then they started to drag us off into these little interrogation cells for some time, and you could hear some heavy duty grunting and moaning going on. And some whacking too. And yelling and screaming because they were going to interrogate us. They were asking me what airfield I come from, and I told them King Abdul Aziz [Saudi Arabia], you know, because they already knew that. There were only so many places you could come from. They asked if they had any planes there and I just made up a number, a huge number. They said how many people at the assault force, I said, "Oh, about half a million, 500,000." I don't even remember now the number I told them, but it was a big number. They went on that for a while. And then they said, "We are going to put you on and show you off." So, they sat me in a chair, pulled my blindfold off. There was a bunch of them standing over there looking at us. One of them started talking, "What unit you in?" "VMO-1." "What plane you fly?" "OV-10." He said, "OB-10" into the TV camera. This went on for a while. He said, "Anything you want to tell your family?" I said, "Yeah, for the children, study your homework and behave your mother." Something like that. He said, "And anything to the country?" I said, "This war shouldn't have happened." Because it shouldn't have. We should have killed that guy [Saddam Hussein] before that.

Anyway, we stayed there for a while, and then they loaded us up into a car and made us lay down on the floorboard, they were afraid someone in Baghdad was keeping tabs on them. Doing some reporting on them. And then they drove us out to this place something like an old prison, it has bars in it and once they took our blindfolds off and we were in our cells, it looked like something outside of an old French Foreign Legion fort. Raggedy and run down. They put us in a cell and one at a time, we were in solitary confinement. They didn't bother us, except for a couple of days, three days. They just left us alone and brought us a little bit of a gruel and a piece of bread in the morning. One day they brought us a piece of meat. And they gave us a little cup of hot tea a couple of times, which was very tasty, you know. It was rich in sugar, so it tasted pretty good.

And then we started to talk back and forth between the cells because in the evenings, all the Iraqi's would leave the compound and lock the big huge gate, the iron gate behind them. And the only people who would be in this prison were us, there were no windows, just two bars. So, we started to talk back and forth, finally we heard the door open and they ran in. "No speak, no speak!" Started beating everyone for a while. "No speak!" Then they left and everyone kept the tone down and tried to whisper back and forth, to see who was there.

We just sat there for two or three days, doing this kind of thing. Nothing else was happening. And then they came in and I thought at the time, if this is all there is too this, then I can do this standing on my head. It's not a big deal. And the way the attacks were rolling in, I sat there and a couple of other guys came to that conclusion. I said, "This ain't going to last three months. They can't take it. You can't keep absorbing damage like that." It was only 45 days instead of three months. Because they were blowing the bejeezus out of those people. But what bothered me was the bombs got closer and closer. I was thinking, "Alright, that's close enough!" "Ah, that's close enough there! You can turn that damn thing around now!" The ground would just shake and shudder from the bombs they were dropping. You could look out the windows and see the air just lit up, like July the 4th. Tracers, a million of them, going through the sky. I kept thinking, "How the hell are they not hitting anything?" Because the sky was full of it. And surface to air missiles, tracers, just going across the sky. They were filling the sky with these things. They had to be firing blind because later I found out that our guys were shutting down all their radars, with missiles or radar jamming. So, they were just launching in the blind, hoping to hit something.

So we stayed there for a few days like I said and then they walked in one night and had kerosene lanterns because they had no electricity. And that's another thing, they had water down at the end of the hall and they gave us this pitcher to keep water in at night. I had one blanket, and it was cold as hell in there. So, in the night we would drink the water and with the same pitcher you would piss in and in the morning, you would take your pitcher down the hall, dump it out, wash it out and fill it back up with water, take it back to the cell.

But they walked in one morning, early, it must have been after midnight, they had a kerosene lantern and they made us all get up, handcuffed us again, blindfolded us and loaded us up on some kind of a bus and then went out to the next prison. This prison everyone called the "Biltmore" because it was huge brick thing, like a hotel. That is where they really started rough interrogation. Really beating on us and sticking guns down your throat, saying they are going to kill you now and pull the trigger—things like that.

They really treated us bad, you know, just bad treatment. For a while there, I was thinking, "I am not going to make it out of here." Because this is getting rough, I thought they were going to beat me to death a couple of times. Thugs, just plain thugs. Kicking you and hitting you.

And I was there and really weak because they were starving us to death. All we had was this little bit of thin gruel and a piece of bread, and that was it for the whole day. My weight started falling off and I looked in my mirror one day and my face looked really bad and I said, "Damn, you look bad man!" I was starting to get thin too. There is a diet that works! No food at all will make you thin. And also, I had two small blankets and I could double one over and lay on the floor but could still feel the brick on the floor. And the other blanket I tried to curl up inside it. I was trying to keep warm because it was very cold in there. It was February. It was cold in those damn cells. I was shaking all the time.

Then I think it was Valentine's Day, February 14th, when they bombed that air raid shelter that the Iraqis [leadership] had all their families in. It was an Intel building so it was fair game. Our forces struck it and a whole slew of them were killed. That set the interrogators off. They came around our cells that night and beat the crap out of you again. They were infuriated.

So, I started to get these sores on my sides and thighs from laying on those floors and on my back for when I was trying to stretch out and these sores started to swell and cracked and they all started popping, I had puss coming out everywhere, on my thighs and my back. I had nothing to wipe it on so I just wiped it on the brick walls. Then my eyes got infected, I had to pry them open every day because all this gooey stuff started to form. Like a thin film of grease and when it was closer to the end of the month around the 24th of February, this is when I stop getting up and walking around. I was running out of energy, I had no strength left. I looked at myself in that mirror and said, "I better start saying my prayers, because I don't think I am going to get out of here." [tearing up] That is what bothered me the most, that Mary will never find out what happened to me. Because they will just throw me out in a ditch somewhere. I will just be someone missing and missing forever. No one will know. That bothered me an awful lot. And another thing that bothered me a lot was on the walls of my cell, someone had scratched a picture, you could tell that this place had been a prison for a long time. The picture was some guy pushing a baby buggy. I would sit there and wonder, "Whatever became of this poor guy." Because I am sure they killed him. So, next to that picture, I carved GH with a heart for loves MH [Guy Hunter loves Mary Hunter].

That day I was being interrogated and the Iraqi interrogator said, "Tomorrow, there will be ten questions. If you have ten good answers you will have ten fingers. If there are ten bad answers, you will have no fingers!" That night I was thinking what driving in the car would be like with no fingers, and turning the pages of a book with numbs. I was thinking, "I can do this." Interestingly enough, I was never interrogated by that Iraqi because our building was bombed that night.

The F-117s [Nighthawk] arrived and started to blow the crap out of that place. I didn't think we were ever going to get out of there until that night. So, it was a weird night, one of those prisoners [LT Jeff Zaun], they blew out his transit window and he got through it and said, "Hey everyone I crawled through my window, everyone crawl through yours and let's get the hell out of here!" He said, "Get out and turn left," and we all said, "Jeff, which way is left?" because the cells were on both sides. I was thinking, "Where the hell are we going to go anyway? We are in the middle of Baghdad. We don't look like Iraqis, not with this yellow jump outfits on that says PW on the back."

So, after that, about an hour I guess, as soon as that last attack, all of us started to yell up and down the hallway. Trying to find out who was there. "What's your name and who are you? What's your unit?" I could hear Acree cry, "Guy are you down there?" I said, "I am down on this end." We all just waited for the guards to come back and when they came back, they were pissed off now. They came back in and started to use pry bars to get the doors open because the bombs just buckled the doors. They started to take us out. When they came to my room, I grabbed my two blankets and I wrapped them around me so tight, I wasn't going to let them take those blankets. Those things were precious to me. They push me back and took those blankets, threw them in the corner and I said, "Well, there goes that attempt." I am going to be cold again.

They took us out one at a time, down the hallway, down the stairs, into the basement, down the basement, up to some stairs to the outside, where it was cold as hell. I had one on either side of me and those flip-flop shoes they had for me fell off so they were dragging me through the rubble and the Brit POW, [Flight Lieutenant] John Nichol. He said, "Guy you are between two guys and it looked like they beat you like Mike Tyson." They were wailing away from you on both sides. And they loaded us into a bus, made us sit on the metal floor and pull the yellow shirts that said PW over our head so we couldn't see around. They made us just sit there and the Iraqi's were walking back and forth with their rifles and every once in a while, they would come over and whack you over the head a couple of times, I was knocked out a few times and dazed pretty bad. We drove for some time on that damn bus and we finally pulled to a stop, they had this huge gate that they opened up. We all filed out and went inside. They made us sit on this courtyard outside in the dark for the rest of the night, until the sun came up. It was cold out there, sitting on that hard surface.

Then they herded us all in a building where they sorted us out. They put all of the Americans in one cell together. I think there was about 13 or 14 of us. We all got into that cell and we were freezing, we all sat there as close as we could, just to try and get a little warmth. Finally, the sun came up and the door opened and the guy came in with a big large pan of gruel in it. I was starving to death. Literally, I was starving to death. And we all said, "Ok, here is what we will do. We will just pass the pan around, everyone take two big swallows and pass it along until we use it all up." That is the only fair way we could think of doing it. So, we started talking and figuring out who and where. The guard came back a few times to say, "No speak, no speak!" But they pretty much left us alone. We found out that the senior most guy there was an Air Force colonel named David Eberly, he was a good guy. He said, "Look, guys, I may be the senior guy here but there is nothing I can do for you. We are all going to be in solitary confinement again, we are not going to be in communications with one another. There is nothing I can do for you. But if you get a chance and get yourself a hold of some soap and water, wash your hands and face. Try to stay as clean as possible because if you get an infection, you are going to die here. Because they are not going to treat you with anything." We all figured that would pretty good advice. Try to get clean if we could. Then we just sat there and waited.

I remember Acree said, "You know what? I think they are going to take us to a farm outside of town, there is going to be milk cows and everything! We're going to eat some good stuff! Just wait until the sun comes up." Sun came up and we looked outside, barbed wire fence, prison walls. Everyone said, "So much for your FARM! And your COWS!" They finally split us up and put us in solitary confinement in this building. That was when "Rat" [Navy Lieutenant Lawrence Slade] got in trouble.

This was a huge complex, just like on the base with building and facilities. It was full of Iraqi's and I found out later on from the Kuwaiti Colonel, these are all friends or relatives or people who Saddam Hussein was suspicious of. He threw their entire families, entire clans in prison, and that is where they all were in that compound, where we were. They had one Iraqi out in the middle of the courtyard. They spent about four or five hours a day just torturing that guy, just pushing him around and around. I could see him outside of my cell window. He was chained to a soccer pole. They were beating him, and shoving him, and pushing him around. They would continue this back and forth—he would be screaming. To tell you how backward their technology was, they had some Iraqi guards out there sitting in chairs while the other guys would get a bucket with soap and razor and shave them. These were the old razors back in the 1950s, the old Gillette safety razor, where you put the single thing in there. That's what they were using to shave each other with.

A couple of days later they opened the cell doors up and put one other person with us. So, we are not in solitary now and I had an Italian in with me, a Tornado pilot. We chit-chatted for a while and they put another fella in with Rat, I don't know who was in there with him. But outside you could see other prisoners with their arms out and there were guards down below us. Rat and his buddy leaned out and waved back at these other prisoners. The Iraqi guards looked up and saw that and it was like the Keystone Cops jumping up. They came tearing in there and busted open that door and started to beat the crap out of Rat and his buddy in there. I started praying for him because it sounded to me like they were going to die right there on the spot they were beating them so bad. But the guards quit after a while, thank goodness. After that is was quiet.

All those prisons, it was deathly quiet. You couldn't hear anything, except people walking around. We stayed in there a few days. The prison was a decrepit old place, you could take chunks of bad cement out of the walls and look through into the next cell. And Acree was next door to me. Then February 28th arrived and the sun came up, that night there was loud noises and shooting and in the morning anti-aircraft was still firing in the sky. The Iraqis were shooting their small arms into the sky. I thought to myself that something had happened, I beat the war is over. A little bit later some Iraqi guards came down, they stopped at Acree's cell and he had this lieutenant colonel Kuwaiti with him. So, the Iraqi guards chatted with him in Arabic, then the guards left and it got awful quiet. I pulled out the chunk of brick on myself and did the ole. (Knock, knock knock, knock, knock) Acree came over and pulled his brick out, he looked through there and I could see him. I said in a whisper, "What's going on?" Remember is the last day of the month, February 28th and he said, "Well, first of all, its pay day!" You got to have a little humor, right? Acree continued, "They told him, the fighting has stopped and this is all going to be over within a couple of days." We actually stayed another week with them.

They let us come down and wash our hands and face downstairs. That was the first time that I had soap and I tried to wash my hands and face the best I could. Then they loaded us up again, took us out to another prison. I don't know where the hell that prison was. They just got through washing out the floors, so there was standing water in mine. I said, "Damn it. I am barefoot, no blankets and now I have water on the floor!" Things are getting worse. Soon enough it started to dry up in there. Some guy opened the door and asked "How many blankets you want?" I said "Five!" So, he just started throwing them in. I thought to myself, "God, I am going to build myself a bed!" So, I stacked them up and stretched out my blankets, three of them in a pile. Did another one for a pillow and then another blanket on top. I said, "Guy, you're in hog heaven now!" I was warm for once in my entire time in captivity.

They fed us a little cheese that evening and some bread—I said "Well, things are looking up here." And then a few days later—we were there a week I guess—the door opened and this Iraqi said,

"Come with me." I went out into the courtyard. There was about 10 or 11 other POWs out there. He said, "You will be going on a ride today. Get on the bus." And then he came around and started offering cigarettes. I said, "I don't smoke." I thought later, maybe I pissed him off because he calls two guys and pointed to me and they took me back to the prison cell. And the other guys got on the bus and went to Jordan. I never did understand that. So, they put me back in the cell and moved me to a different cell. I was sitting there for a couple more days and I was thinking they aren't going to stop letting people go, they will let the rest of us out eventually. I was in pretty good spirits about that. A couple of days later the door opened and the Iraqi said, "I can tell you this, you are going home to your country today." They came by, made us go down to the end of the hall, take off our uniform and a big pot of warm water, with a washcloth and some soap. So, I got to take buckets of water and wash and rinse myself off.

Then they gave me a brand-new POW outfit and then they gave me some shoes that didn't fit, tennis shoes. Took me back to my cell. I was glad to have some shoes now because that floor was cold all the time. Back in the cell, I waited a while and finally, the cell door opened and he put a blindfold on me, lead us down to the of the hall, they opened another door. We walked out and I could tell we were going toward some type of vehicle because we bumped up against it and we had to step up into the bus. As we got on the bus, some guy next to us was spraying us with some type of cheap ass perfume. Spraying it on so we wouldn't smell so bad I guess.

So, we got on the bus and started to move and they took off our blindfolds and took off our handcuffs. I said, "This is looking up!" Guards were there with their AK-74s and the guy next to me, we started chatting and I said, "By the way, what's your name?" And he said, "Don't you remember me? Colonel Eberly!" I said, "Jesus Christ Colonel, you look like shit!" He looked like he was about to die, right there on the spot. He was a Strike Eagle pilot, F-15D. Anyway, we drove on this bus for a while and we saw some other guys we knew up front, Harrier guys and other folks. I was waving at them. We pull into this courtyard and there was a hotel, and they had armed guards around with machine guns. They said, "We are going to stay here because we can't take you out of the country today. There is a sandstorm moving in and the Iraqis don't have complete control of their anti-aircraft units around town. So, we cannot fly in Red Cross planes to get you. We will stay the night and with luck, we will leave tomorrow." So, we stayed there the night, and they said, "We will have electricity part of the time, and we have generators, so you will have some warm water to bathe for some of the time."

"We don't have enough toothbrushes to go around, but we do have some soap. So here is a toothbrush, one for every three to four guys. And here is a tube of toothpaste." So, we went and brushed our teeth, passed the toothbrush around. Then they put us two to a room, we were all in one floor. At the end of each floor were machine gun positions facing outboard, the Red Cross hired these guys just in case the republican fanatical guards were coming back, the machine gunners were going to shoot them down. The Red Cross guy said, "This is here to protect you! Not to keep you in." So, we got in the room and while the power was still on, we all started taking showers in our own separate rooms. I took a shower and washed up as hard as I could. And Acree came out and said, "How do I look?" I said, "You need to go back in there, you have filth and grime ground into your head!" So, he went back in and took another shower, tried to get as clean as possible.

Then the Red Cross came around and gave us a few pieces of chocolate, that's all they had for food. But you know, I was happy to get that, happy for that. The power went off, and everyone decided it was time to get into bed and sleep some. We each had two separate beds, but it was just too damn cold, so we pulled the beds side by side, threw both of the blankets from the beds on top of us. I just passed out, went to sleep. I was dead tired. We got up in the morning. The Red Cross had some more chocolates for us. A little while later, they said, "We are going to leave now. We are going out the side door. People are gathered around outside and a lot of them are Iraqis, and a lot of them don't like you. They would like to see you dead so don't look at them. Don't make any hand gestures to them. Don't wink at them. Don't do anything. Just get on the bus and sit there looking straight ahead." So that is what we did. We got on the bus and traveled through town and boy you could see the damage those bombers had done. There was a communication complex that was utterly destroyed. Next to it, the buildings were ok. Some very accurate bombing they were doing.

We got to the airport and as we got there some planes were landing and they were bringing some Iraqi prisoners back. Bringing them back to Iraq and everyone was looking at them getting off the planes and we said, "These guys don't look happy at all to be back here!" They look like they were dreading to be back there, to Baghdad. I said, "I bet they felt sorry they ever got on that plane to come back here." So, they finally brought a Red Cross C-9 [Nightingale] in and we got onboard. It was a Swiss airplane. And finally, we took off and just as we took off and broke free from the thin scum layer, we were joined up on both sides with some F-15s. [tearing up and shaky voice]. It was really good. [wiping his eyes]

RETURNING HOME

And then we started heading out, and they said, "Welcome back to America!" We flew into Riyadh, walked off the plane and there was Schwarzkopf and all the big wigs. We got onto another plane and went to Bahrain, got off there went on a bus to a hospital ship. We stayed there for two or three days, may have been four days or seven, I don't remember. They brought us all our uniforms and caps, flight suits and boots. We didn't have any money, you know. I wanted a watch in the worst way to see what time it was. I borrowed $25 dollars from our doctor and bought myself a $15 dollar watch so I could tell what time it was. We got on a plane and they said, "We are going to fly to the islands off the coast of Africa." Canary Islands, I think? So I went to sleep, and I woke up and we were descending, and it's as green as hell around there. The guy next to me said, "We are going to a different route. We are landing in Ireland." So, we landed in Ireland and as we got off the plane, the doctor said to all the POWs, "I know we are stopping in Ireland, all your escorts, they can have something to drink at the bar at the airport terminal, but you guys can't because you all have some type of parasite and the medicine you have been taking, it will make you deathly sick. So, don't drink." So, we said, "Alright!"

We stayed in Ireland at the terminal for a while and then we got back on the plane and flew to Andrews AFB [Maryland, near Washington D.C.]. At Andrews, they put us up at Bethesda, all the Marines and Sailors, at Bethesda Naval Hospital. They hooked us up, but unbeknownst to us at the time, Bill Marriott, head of the Marriott Hotel Chain said, "I want all the POWs to stay at my hotel for free. Everything is on me, for as long as they want for a month. Everything is on me, anything they want to order." When we were there, we had a reception at the hangar at Andrews, I got to see the Commandant again. He said, "Well, Guy it's good to see you again." I had known him for years and here he is a four-star general. I had known him since he was a full colonel. I had talked with him back and forth. He said, "Guy is there anything I can do for you?" I said, "Sir, I would really like to stay in the service until the OV-10s go away!" He said, "Well, the OV-10s will be here as long as I am Commandant!" I said, "I am coming up on my 30th year. I would like to stay in a little bit longer in the Marine Corps if I can!" He said, "I will take care of that!"

I tell you, I got back and later on the OV-10s did stay around as long as he was Commandant. And then I hit my 30th year, and I had orders already. We were going to drive to the east coast and I was going to retire there, in Jacksonville [North Carolina].

Did you have trouble sleeping and flashbacks of your captivity?

I had trouble sleeping and would have nightmares for years. I would wake up screaming and yelling. Mary would have to shake me to wake me up. I would be sleeping and I would swear to God I could hear them [the Iraqi guards] walking down the hallway, I could hear the steps and the keys jingling. Because if keys are jingling, someone is being taken out for interrogation. Those kinds of dreams bothered me a lot. I would scream out. During my Vietnam experience, I never had flashbacks, but with my experience as a POW, I have dreams of Vietnam as well. I am going on mission and have missions to plan. Vietnam was my war, my generation's war.

LEADERSHIP

During your captivity, was there any leadership you witnessed?

Everyone was looking after everyone. Specifically, Colonel Eberly's comment. He said, "Look guys, I may be the senior guy here but there is nothing I can do for you. We are all going to be in solitary confinement again, we are not going to be in communications with one another. There is nothing I can do for you. But, if you get a chance and get yourself a hold of some soap and water, wash your hands and face. Try to stay as clean as possible because if you get an infection, you are going to die here. Because they are not going to treat you with anything." We all figured that was pretty good advice.

Looking back on your 32 years of military experience, what leaders did you serve under that you felt had the right stuff?

There were several good COs in my life, in fact, there were only two COs that I didn't like. But overall, it's been my experience that if you get to be a lieutenant colonel and you get to be a squadron CO, you are pretty good at being a people person. You can handle people. I had several who were really good. Colonel John Fridell was really good. I did a paper on him when I was in college. He looked like a Kansas farmer. He was tall, slender, wrinkly face. He ambles when he walks. He was a good guy. I emulated him in a lot of ways. He was always calm. He was always upbeat. He was always trying to help his officers out. This is a man who likes his officers in his battalion. People liked being around him. He had been a rocket battery commander in the Korean War. Before the intercoms had been installed, he had bought his own intercom and installed it at the office. He would use the intercom, "Hunter, why don't you stop on in here [his office] in a minute?" I would go into his office, and he was just bored. [laughter] He would move his couch out from the wall and pull out this footlocker, and he would start showing me pictures of everybody and what he did in Korea. Talking about missions they were on and shooting rockets. He had a really good sense of humor. I really liked him a lot and really respected him.

What advice do you have for officers who are currently serving?

First off, you need to know your job. If you're a pilot, know what the hell you're supposed to be doing and how to fly that plane, and fly it aggressively. There is an old saying, "Fly that plane to the edge of the envelope and not one second past it."

Also, be firm, be fair, and be consistent. If you are going to be nasty, be nasty all the time because people will know what to expect. But you still need to be firm and fair. People don't mind nastiness as long as you are an even keel person. And you will be treated fairly. That's about it as far as I am concerned. There is nothing secret about it. Just try to be an officer who can see the other persons view point.

Guy Hunter, Jr. experienced something many will never truly comprehend. It was pure hell for those who were captives under a desperate and inhumane Iraqi military. What makes this ordeal even more profound is that he, along with his pilot Lt Col Cliff Acree, demonstrated uncommon valor in flying the missions leading up to the day they were shot down. At the start of the Gulf War, the Iraqi military was a capable force and maintained weapons able to inflict damage to United States forces. Aerial observers, along with the aircrew that fly their missions, are a noble bunch, putting themselves directly into harm's way in order to protect friendly forces and destroy the enemy's ability to fight. Guy Hunter, Jr., on behalf of a grateful nation, "Thank you for your sacrifice."

■ Interviewed by LCDR R. Thomas Griffin, US Navy, and Maj Fred Merrill III, US Air Force.

CWO Guy Hunter standing next to OV-10 Bronco.

CWO Guy Hunter sitting in the OV-10 Bronco.

CWO Guy Hunter's OV-10 Bronco crash site.

CWO Guy Hunter with fellow POWs released.

CWO Guy Hunter
POW video while
in captivity.

Guy and Mary Hunter are reunited after his release.

UNITED STATES
TRANSPORTATION COMMAND

Walter Kross

DIRECTOR OF OPERATIONS AND LOGISTICS, TRANSPORTATION COMMAND

General (ret) Walter Kross is a United States Air Force Combat Veteran whose military experience, from 1964 to 1998, was quite extensive, culminating as Commanderin-Chief of the United States Transportation Command and Commander, Air Mobility Command, Scott AFB , Illinois. As a unified Commander-in-Chief, he was responsible to the SECDEF for the Nation's defense transportation requirements. He exercised command over service transportation components from the Army, Navy, Marines, and Air Force. As the TRANSCOM Commander, he oversaw a 168,000-person organization with more than 1,500 aircraft, 120 ships, 1,400 specialized railcars, 12,000 containers, a worldwide network of aerial ports and seaports, and the corresponding physical and information technology (IT) infrastructure to operate daily across the entire world. Gen Kross is best known for his vision, guidance, and contributions as Director of Operations and Logistics for all defense transportation requirements in Operations DESERT SHIELD and DESERT STORM. As the Director, he orchestrated the largest transportation operation since WWII, shipping 3.5 million short tons of cargo, along with 6.1 million tons of Petroleum, Oil and Lubricants (POL), and over 503,000 personnel. Gen Kross was the first in his family to attend college and he was commissioned through USAF OTS in December 1964. His early career combined both fighter and airlift experience as he flew 157 F-4 [Phantom] combat missions, with 100 over North Vietnam. He later transitioned to strategic airlift, followed by a tour in the 89th Wing at Andrews AFB, Maryland. He was assigned to the Air Staff for six years, including 2.5 years in the Chief's Staff Group as the Theater War subject matter expert. He served as Wing Commander of Dover's C-5 [Galaxy] wing, USAF's Director of Operations Provisional Commander, standing up Air Mobility Command (AMC), then served as AMC's first Vice Commander. He was Commander of 15th Air Force and Director of the Joint Staff in the Pentagon. He was an instructor and evaluator pilot with almost 6,000 flying hours in 12 different aircraft ranging from the F-4 to C-141 [Starlifter] to the KC-10 [Extender] and the VC-137 [Stratoliner]. In Vietnam, he earned three Distinguished Flying Crosses and 13 Air Medals. He founded USAF's Project Warrior Program and produced the USAF's Trainer Aircraft Master Plan. He conceived and championed the DOD's important Global Air Traffic Management (GATM) program, teaming with the Federal Aviation Administration (FAA) to establish and fund what has become a multi-decade program of continuous aircraft avionics enhancements, enabling safe and efficient air operations worldwide. Winston Churchill stated, "We make a living by what we get, we make a life by what we give." General Kross sacrificed for our country for 34 years. His leadership, life, and legacy will continue to serve as an inspiration and a tool that shapes tomorrow's Air Force.

We would like to capture your days before you were commissioned in the Air Force. Can you tell us about your childhood and upbringing?

My parents grew up, met, married, and raised four boys, all in New York City, specifically the Bronx. My father was Polish and Belarusian. My mother was Irish, through and through, with red hair and freckles—the whole works. All my grandparents immigrated to America in the early 1900s, each having a classic and harrowing journey our family is quite proud of. My paternal grandfather journeyed as he escaped the Russian draft so he didn't have to serve on the front lines of WWI. He entered illegally, crossing a railroad bridge over the Niagara River Gorge one winter night with my grandmother and baby. My Irish grandparents came through Ellis Island separately. So my parents were first generation Americans. We were a blue-collar family.

We lived in the Bronx until I was about 12 and then we moved to Bethpage, on Long Island. Like a lot of people, when your parents had saved just enough money, my dad bought a house out in Nassau County, Long Island—it was the 1950s migration to the suburbs and the pursuit of a better life. I graduated from Bethpage High School where I was starting center on the basketball team.

My dad was a member of the "Greatest Generation." During WWII, he enlisted in the Army. He was awarded two Bronze Stars and a Purple Heart. He was an infantry squad leader and went all across Europe right to Victory in Europe Day (VE-Day). All four of us boys grew up asking our dad for bedtime stories about his experiences in the war. So military service was always in my mind, but not overtly. I never thought about going to the Air Force Academy. Never would have qualified. It was beyond my vision. The Air Force Academy was a new thing then, a bright shiny penny too far west for a New York boy. Nor was I one of those guys who always wanted to be a pilot or was the kid that was always looking up in the sky, yearning to be a pilot. I was just a regular kid growing up, concentrating on the next normal thing. College. A good job.

I went to Niagara University [New York], majored in chemistry the whole time, and graduated with a B.S. in Chemistry. It's a land grant college, so it was mandatory for all male students to be in ROTC for the first two years. We learned how to field strip M-1s and march around, but I had no interest in continuing after the mandatory part. I played varsity golf all four years, captaining the team my senior year, so I was more interested in balancing chemistry lab and intercollegiate golf. After graduation, I was accepted at St. John's University [New York] to get a master's degree in chemistry.

That summer of 1964, I worked at the World's Fair in Flushing Meadows [New York] and one day I was giving a coworker friend a ride home. We made a stop at the Air Force recruiter on the way home. I waited for him. After having some pizza and beer across the street, I went looking for him. He was inside taking the Air Force entrance test. So, I took the test too, just to kill time while waiting for him to finish. It was the Air Force Officer Qualifying Test (AFOQT). Three weeks later a postcard came in the mail. On the back, they offered me an OTS slot, followed by pilot training. On a postcard. The Vietnam War was expanding and flying slots came easier then. So, just a walkup test, followed by a postcard.

Back then, every young male was confronted with how to deal with Vietnam and the draft. That postcard laid out my path. Jet pilot sounded much better than grad school. I joined more as an alternative choice to the inevitable Vietnam draft.

When you go to OTS, they, of course, are trying to take raw material and validate the AFOQT tests. "Can you really be an officer?" If there was no Vietnam War, I would now be a retired chemist. So, this is the autumn of 1964, and OTS was shortened to 84 calendar days for Christmas. Halfway through OTS, the active duty officer overseeing your flight of cadets gives you his midterm assessment. Then Captain George Fox calls me in and says, "It is questionable whether you can be an officer. You don't really have officer bearing." Everyone in the lower half of officer training school is an officer trainee second lieutenant. One bar. When you advance to the upper class six weeks later, they give you a rank all the way up to cadet wing commander, depending on how you did during the lower half. The lowest possible rank you can receive is cadet first lieutenant. That's what I got. Everyone else in our flight of 18 was promoted to officer training (OT) captains, majors, even lieutenant colonel . They were heads of the parades or wing staff. I was given absolutely no responsibility except to graduate. They saw no potential in me, and they were right.

We are all laughing at this point because we are talking to a retired Air Force four star general, one of the highest ranking officers in the military.

As you were transitioning there, you talked about how Capt George Fox didn't feel that you had the potential to become an officer. What were your thoughts then? For instance, did you question yourself such as, "Did I choose the right path?"

I knew he was being truthful based on what he saw of me. When he talked to me about my officer bearing, I thought, "Well, ok, I am a big lanky, skinny guy." So, I took it as valuable feedback, certainly adjustable. "Well, I will stand up straighter. I will run faster in my physical fitness test, and I'll try to do what he wants me to do." By this time, I wanted that officer's commission. I took it as a challenge to do better. I got a little bit better in the second half of OTS, and that kind of continued for the rest of my career such as it was. I remember Project X at OTS. It was all about being a valuable member of the team and helping the team accomplish the mission. So they were telling me the playbook, the script. Not rocket science compared to chemistry and certainly crystal clear, I got it. "Improve. Stay in the here and now. Help the team do this mission."

I got a little better in pilot training, finishing sixth, neither here nor there, but respectable. And so that was the formula for me, continuous improvement all the time. Not obsessively, but I felt that as I went through the Air Force all the way to my last assignment, that I was continually challenged to do the mission, to help the team do the mission; in each and every assignment.

As we continue to be challenged, to include currently, in ACSC, your wisdom is priceless. I think what you are saying is to stay engaged, keep working hard, and keep doing the things that you need to do to, include working on weaknesses and to continue to improve because it will ultimately pay off, as it has done for you.

You will hear me all morning using certain terms and continuously going back to them. One dominant theme: "No one knows about oneself until truly stressed." It is simple human nature. It is just the way it is. This will not change, ever.

LEADERSHIP

What would you consider to be the most important leadership trait that you possessed?

That is a very good question; I have three. Three levels of abstraction, which I believe is truly important, because an effective leader has to operate on all three levels.

This, first and foremost, is listening. A tool? A technique? Or an ability? I say the latter. Listening is critical. As you age, you are always surrounded by people from whom you are generationally separated. Young people who have the talent, commitment, and dedication to see things through a different lens. You have got to be able to take their strengths on board and factor them in. It doesn't matter about rank. Some of the best ideas are from the American enlisted force. If you transmit that you have all the answers, then you are doomed. And, worse, so is your mission, since you alone are of no real consequence in and of yourself. There is one small little tool you can use for listening that can be very invaluable, wait for the period at the end of the sentence. A little more on listening. It's a useable tool, to be sure. As a technique, wait for the period at the end of the sentence. Watch how others don't sometimes. Someone else will be talking and they will cut off those last couple of words. You may have cut off their concluding thought or how they may wrap up that thought into something even more valuable. Listen for the period in the sentence. Now that will cause you to be sometimes at a disadvantage because someone else will chime in because they won't wait for the period. This is particularly important with one-on-one conversations or almost always with people with less rank than you. They are looking at things very differently than you are and if you cut them off the first time, they might not continue a second time. You have to really give them the sense that what they are saying and doing is important.

The second one is the absence of a negative trait. It is to be totally without guile to the point of outward self-deprecation. You cannot be an effective American military leader with a large ego. Self-assurance, sure. Inner strength, definitely. Knowing one's place in the universe, one's place in space and time. Clearly, an acute and heightened sense of situational awareness flows from managing one's guile. Said another way, "Be Humble. Always. In all things."

The third is honor. The highest level of abstraction. Honor in all of its manifestations. Honor almost to a fault. We expect it so much from everyone in our profession that we almost take it for granted. So much constantly flows from honor, ethics, standards, even the other two it is paired with, duty and country. They really cannot stand alone without honor.

So there they are: listening, lack of guile, honor.

Are leaders born or are they made? What is your perspective?

Most are made, but you have to start with the right material; a combination of both. You have to have certain personality traits in your DNA, in your behavior, in your temperament. Start with a blank canvas upon which all the music is written and eventually make yourself a finished work. You are not born a finished work. May I ask you a question back? "Do you think a good leader has to be a good communicator?"

It is an important trait, but I would say it depends. Sometimes actions speak louder than words. As I've learned from our Marine counterpart in our ACSC class, there are some things the Marines do in their culture that I really like and I wish the Air Force would take advantage of. From the standpoint of a leader, if you're the highest ranking, you're the last one to eat. In the Marines, you make sure everyone in your company is fed and taken care of before you take care of yourself. I think that in certain aspects, just by showing that you care, people will go to the ends of the earth and work for you. They will try their hardest and work night and day for you if they know what they're doing matters. In addition, you give them the true thank you by saying, "I really appreciate everything you've done for me/squadron/unit." I think that's very well said. And of course, in our world, we go to a lot of dinners. You'll be going to more dinners in life. The perfect example of that is that your head table should always go last through the buffet line or your table gets served last from the kitchen.

But on the communication point, as you know, being a good communicator has to be a necessary skill in many cultures and many environments. But you also know that there are exceptions to that rule and there are times when you will follow someone to the ends of the earth, even if he's not a great communicator. But he demonstrates through his actions, through his persona, that this man is the right man at the right time and I need to follow him. He may forego communication at times or he may just be a guy who is really not a good communicator and you almost have to draw it out of him, but you follow him because of his actions and his character; his honor. Which again proves, it's not a technical toolkit, it's an art form.

What experience helped shape and develop you throughout your career?

One event set me on a course that changed everything; the clipboard incident. When I graduated from pilot training, everyone from our pilot training class were assigned as F-4 backseaters or as C-130 co-pilots. While I was at Laredo [AFB], Texas, one of the "new" F-4s actually flew in. God knows why. I was able to sit in the front cockpit. It was like sitting in a living room. It was so roomy. It was the first fighter that

I actually fit inside of. Anyway, I was sent to Davis-Monthan AFB [Tucson, Arizona] for F-4 initial training as a backseater. There you married up with a front seater, and you went through the training as a paired crew. My official assignment was to Bitburg AB [Germany] as a backseater in the F-4 in United States Air Forces Europe (USAFE). Our front seat aircraft commanders were from Bitburg already, converting from F-84s [Thunderjet] to F-4s.

One day we were ordered to go to the base movie theater, and there in the front of the room was the Wing DO, Col Daniel "Chappie" James. All the people in the front row were host wing instructors. Col Chappie James says, "A bunch of assignments have come down. We are at war. We are recasting everybody, but we are looking for volunteers first." The assignments that came down were meant to, upon course completion, redirect F-4 pilots to Eglin AFB [Florida] in order to build four combat ready squadrons there, then deploy them as units to Southeast Asia.

That day in the base theater, Chappie James said, "I got this clipboard here and I will be passing it around. We are looking for volunteers to go to the war." On the spot, he was asking us to give up this plum assignment in USAFE. There were five of us backseaters, all bachelors, seated in the last row of the theater. When the clipboard got back to us, there was not a single name on it, and that's despite the fire and brimstone speech by Col Chappie James. Five of us decided to put our names on the clipboard, and our lives careened off into uncharted space. We were the "Clipboard Five." Four were in our wedding eight months later, in a place I never thought I would be.

Our assignments to Bitburg, Germany were cancelled. We were assigned to Eglin AFB under the 33rd Tac Fighter Wing Commander there, Colonel David C. Jones. He later became the Chairman of the Joint Chief of Staff. We were sent to marry up with experienced front seaters there, and train to become combat ready in the first deploying unit, the 25th TFS. There were four squadrons in our building. The building is still there. Cinder block is forever.

I was paired up by chance with a hot shot F-4 front seater at the time, Captain Mike Loh, who eventually became the Air Combat Command Commander. We are still very good friends today. You asked me who influenced me—I watched this man in his prime—in his element. He was a completed work, right there, just for me. So I went to school. He was pure gold as an officer and an extraordinary fighter pilot. I remember he would say to me, "Walt, you'll never make captain if you continue to think like a lieutenant." Simple, yet profound at the same time. Getting assigned as his backseater was a priceless gift that even I realized I should not squander. He was so well respected early on that the Host test wing at Eglin noticed him. The test wing was doing a lot of tests on the Sparrow radar missile and Sidewinder infrared missile. The Test Wing tapped Capt Loh and his pitiful backseater to go up and shoot down QF-104 target drones. We got platinum experience. QF-104s were the size of MiGs, How good was that? We were actually shooting down real airplanes while testing the envelopes of those missiles. We were shooting at 1.7 Mach in trail. We were shooting 1.7 Mach head on. Do the math. Head on, we had to lock on at 50 miles and then break immediately after missile release so you didn't fly through the debris. I was getting a leadership laboratory and building priceless functional skills at the same time. Talk about pure luck in a turbulent world.

The turbulent world continued. It seemed there were disruptions happening all the time. One hit squarely on the Clipboard Five. The F-4 replacement training unit at George AFB [Victorville, California] had immediate F-4 front-seater slots that needed filling in two weeks. Again, the Clipboard Five got tapped for these slots. So, I left Captain Mike Loh after six months to upgrade to F-4 aircraft commander en route to Southeast Asia for combat duty. Four of us survived our combat tours. Shows what forks in the road can do in this life. We all have them in our lives. In my case, I took a chance and volunteered. Fate and timing played massive roles.

Arriving at Da Nang AB [Vietnam], I checked into the 390th TFS—the Wild Boars. I was the youngest F-4 aircraft commander. That means permanent wingman. The quintessential Blue Four, or really Blue Two most of the time. Remember Captain George Fox? Well, there he was. During my combat tour, I flew his wing about 20 missions over North Vietnam. He later told me, "I never thought this was going to happen." Frankly, neither did I.

When I was coming out of Vietnam, I had flown 99 missions over North Vietnam as a first lieutenant aircraft commander. I pinned on the day before my final mission, so I flew one mission as a captain. They let me lead the flight on that last mission. Still remember the call sign: Gunfighter 01. Felt good. However, I was still a first lieutenant when they were doing up follow-on assignments months earlier. They were not going to put a first lieutenant back into the replacement training unit as an instructor pilot. So, Personnel said, "You are getting your geographic region of choice, not your airplane. You are going to Charleston [South Carolina] to fly C-141s. You are one of the lucky ones." So I said, "That's ok. I'm good." But, initially, inside, I felt like this was not so good. I was going to an aircraft and mission that I didn't want to do.

I was dead wrong. It was the best thing that ever happened to me. Fate's hand once again. I kind of moped around being a C-141 co-pilot, learning this new crew culture reluctantly, and keeping it to myself. Then bingo, when I upgraded to aircraft commander a light came on and everything changed. Every mission was a leadership laboratory in itself. You are put with completely different crewmembers you have never flown with before. We flew all around the world. Literally everywhere. It was leadership heaven. Charleston was a place that was not repetitive; you did everything, from tactical airdrop missions to round-the-world embassy missions. You flew everywhere; to include all of Vietnam, South Africa, and South America, even Greenland and Berlin, Germany. Truly all over the place. I was getting all this phenomenal training in strategic airlift, building lots of flying time, plus leading crews on every mission. We dealt with everything. I dealt with absent without leaves (AWOL), customs cheaters, unsafe crewmembers, even closet alcoholics. This was still the time when people had high balls at five o'clock every day. I remember opening the aircraft latrine and finding a crewmember downing a flask and dealing with this as an aircraft commander. On the road. It happens when you are in Johannesburg [South Africa] or at 35,000 feet, and you have to put your leadership skills into practice.

Late in the Charleston tour, I was tapped for Squadron Officer School (SOS). Pound for pound, dollar for dollar, day for day, that was absolutely the best professional military education that I have ever received. It was a robust fourteen week course then. Personally, as a seven year captain, it was perfectly timed for me, developmentwise. There were 800 students in our class. I got to SOS and it was yet another superb leadership experience within itself. Whenever you do anything in the Air Force where you put everyone from all different disciplines all in one crucible, there is an opportunity to see who and what bubbles to the top. This produces tangible, professional capital that is a gift or talisman that keeps on giving one way or another for the rest of one's career. The stamp on one's training form that says Distinguished Graduate (DG) or Top One Third—that is "professional capital." Like other circumstances it differentiates you in your professional culture now. It becomes part of your permanent record. It is predictive in that it validates other elements.

SOS was teamwork and individual performance, concentrated professional learning, spoonfed. It was not about doing it alone. You do it as part of a team and you use leadership to create performance within your group. And you learn all the little tactical things like: listen for the period at the end of the sentence; don't shake change in your pocket; have five main points; and have a beginning and end. It is all the things that I learned at SOS that I applied for the rest of my Air Force career and beyond to great advantage. I will never forget those SOS lessons.

My tour was ending at Charleston and my wife, Kay, said, "What do you want to do next?" I didn't have the Vietnam pressures anymore. Kay reads everything. She read the *Air Force Times* cover to cover. So she says, "The Presidential Wing at Andrews AFB is looking for pilots. They had just lowered the entry requirements from 4,000 to 3,000 hours. I had just passed 3,000 hours as a captain. I ended up being a VC-135/137 [Stratolifter/Stratoliner] dual-qualified pilot and aircraft commander for the next 2.5 years. Again, it was another form of leadership heaven to me. The 98th Marine Air Support Squadron scheduler would give me a name of a very important person (VIP) staff contact on a 3x5 card and then you plan, conceive, and equip the mission right down to the meals, visas, and everything else. Then you would fly and execute all elements of it. Some of the missions were 35 days with 35 stops. My wife, Kay, would follow us on the nightly news. Got to fly the historic [Secretary of State Henry] Kissinger Middle East Shuttles. [He engaged in "shuttle diplomacy," which was various short flights among Middle East capitals.] I remember being the first crew into Damascus [Syria] since WWII and commanding a round-theworld trip, carrying the Apollo 17 astronauts on a diplomatic mission. Even an Air Force One and an Air Force Two mission.

Near the end of my tour at the 89th, I was promoted below the zone to major, and sent to ACSC. Part of the reason I was promoted early was my performance in fighters and in the airlift world, further validated by DG at SOS. I finished in the 99th percentile in academics and overall at SOS as a DG. I do not understand that. I am no academic whiz. I just studied the coursework. It had not happened before in the 89th. In the end, I was promoted below the zone to major, lieutenant colonel, and colonel, but just one year early each time. It turns out that three times early is a rather rare thing. Nonetheless, when I was selected for brigadier, I was the second most junior, even with a total of three one-year below the zones earlier. The Air Force appeared to be looking for sustained superior performance in selecting their new generals.

Leaving the 89th, I went to Maxwell [AFB, Alabama] for ACSC. It had only been 2.5 years since SOS. ACSC was like a one-year drawn out SOS. Fourteen weeks spread out over a one-year course. So, I used it for professional reinforcement. The course content was generally the same, and the speakers were generally the same. But again, it was all about validating, and going deeper into a lot of things that I found helpful and useful. And I picked up a second master's degree. When I graduated from there, they didn't have stratification or percentiles, just the DG program. I graduated as a DG from ACSC. So, by the time I am going into the Pentagon, I have gone from a marginal OTS cadet to a DG from both SOS and ACSC just through continuous improvement and focusing on the mission and the team. I hit my stride professionally by following these mantras: listening carefully; staying humble; staying honorable; not looking over the horizon; doing the mission currently assigned; doing the job; making the most out of it; and volunteering a few times when it is wise.

Taking your 34 years of experience, what would you tell Second Lieutenant Kross?

Listen; stay humble in all situations; be honorable; and continuously improve as a member of the team, focusing on the mission.

DESERT STORM

When the US entered DESERT SHIELD/DESERT STORM, as a leader in the position you were in, what came to mind when you learned that we were going to war in Iraq?

Well, first I need to set the stage and tell you a little about TRANSCOM and

CENTCOM for a front-end transportation perspective. Of course, the mobility war started in the first week of August 1990 for the US—DESERT SHIELD. The hot war started on our side, of course, later as DESERT STORM. TRANSCOM was only two and a half years old—a mere fledgling—still feeling its way. It was formed out of the Goldwater-Nichols [Reorganization] Act of 1986, which cleared the way to establish a consolidated joint transportation command. If it were not for the Chairman of the Joint Chiefs of Staff, Admiral William J. Crowe, who cast the deciding vote 3-2 in the JCS Tank, we would not have a TRANSCOM. The Marines and the Navy voted against it, and he, as an Admiral, voted for it—true jointness—wisely done. Nonetheless, when the DOD activated TRANSCOM on 1 October 1987, it was merely "half a loaf." It only had wartime authorities. It had no peacetime responsibilities except for joint transportation planning. Remember, no organization knows much about itself until it's truly stressed and that certainly applies. There are a number of uncontested truths in our US military world. These truths have the status of immutable physical laws. Violate them at your peril. DOD and the JCS did just that. When standing up TRANSCOM, they violated one immutable truth, "Train as though you will fight."

As a result of violating this golden rule, no one really knew just how unprepared the DOD was for its first mass-sized worldwide deployment. Not the JCS, not the Combat Commands, not the services, not even TRANSCOM itself. TRANSCOM's limited peacetime authorities really created a false sense of security. It could participate in exercises and could build Time Phased Force Deployment Data (TPFDD), but it didn't have any peacetime authority to affect operations and contracting of its own service component commands: Military Airlift Command (MAC), Military Sealift Command (MSC), and Military Traffic Management Command (MTMC), and the Defense Courier Service (D.C.S). So for its first 2.5 years of existence, it had no real control over its own components. Consequently, its own components, as well as the regional joint commands paid it little regard, and that meant that TRANSCOM had never stressed itself in any exercise of any kind to do its wartime job. It was a toothless nothing, up to that point. It had a great commander from the outset, General Dwayne H. Cassidy, himself dualhatted as the commander of the MAC. He handed over TRANSCOM to General H.T. Johnson several months before Iraq's invasion of Kuwait. TRANSCOM's most experienced senior transportation officer, Maj Gen Jack Griffith retired after 42 years of expert service, one week before Iraq invaded Kuwait. Sadly, I was his replacement.

We had some good fortune in that CENTCOM had just completed an operational plan (OPLAN) that was based on a reasonably accurate scenario, whereby Iraq invaded Kuwait and Saudi Arabia. OPLAN 1002-90 was the war plan. They actually got the scenario somewhat right: Iraq invades Kuwait, but then Iraq keeps going for 1,200 miles into Saudi Arabia. OPLAN 1002-90 assumed 50 percent casualties before coalition forces stopped the Iraqis and pushed them out of the invaded territories. CENTCOM even conducted an exercise [INTERNAL LOOK] in July 1990 to assess its OPLAN. It identified serious deployment shortfalls in surge lift of all kinds. Nonetheless, this OPLAN essentially became the playbook for DESERT SHIELD and DESERT STORM, but it was no real value for DESERT SHIELD. That's DESERT SHIELD with a small "d" and a small "s," because the OPLAN just assumed away transportation. You see, for an OPLAN to be complete and executable, it must be declared "transportation feasible." Officially, CENTCOM just assumed that they would have 30 days warning, 20 days of advance notice, before the enemy came across the line into Kuwait, and 10 more days before defending the Saudi oilfields.

Planners just assumed everything was going to be there. In 30 days, they would have two armored divisions, a Marine Expeditionary Force (MEF), 28 fighter squadrons, all the players, all the tanker support in place. But, with no official transportation feasibility to this plan, that meant there was no valid refined TPFDD, the master movement plan for deployment in full third-level detail. There was a TPFD but it was not the product of a refined TPFDD conference that would make it executable. It was just a generic TPFDD. When the war started, the TPFDD contained many unit line numbers (ULN) for equipment that had already been taken out of the inventory. It did not contain some ULNs for new systems that had come in, and the priority sequence in which units would mobilize and move was all from the very outset. Lots of changes were needed from the outset.

In the 1980s, joint planners knew all these problems existed and needed to be solved. The big bet was an overarching system and process called Joint Operation Planning and Execution System (JOPES). But, JOPES, the supporting technology, and even the properly trained professional cadre necessary to make JOPES work for the warfighter was really ten years away. JOPES was just really rolling out. We had a few JOPES-based exercises, but we had never deployed or fought a real conflict using it. JOPES-trained people were like hen's teeth. Most of the people who were critical to the front-end deployment surge were not trained in JOPES operations, processes, and procedures, and JOPES processes were largely immature, JOPES required the rigid Worldwide Military Command and Control System (WWMCCS) system and its dedicated computer terminals. None of it had ever been truly stressed.

WWMCCS command, control, and communications (C3) was a real handicap. This was still the age of WWMCCS terminals. This was a dedicated DOD C3 system. It required having specialized WWMCCS terminals and just as important, software systems to run on WWMCCS terminals. People had walked through JOPES for exercises through small little things, but it was highly orchestrated and highly synchronized. The people who walked JOPES were JOPES people, and they would be set to whatever exercise and operate the WWMCCS terminals. They would load a ship or load three aircraft after a fashion, and then they'd write up lesson learned. But, to do that on a massive scale—well, almost impossible.

But, the die was cast for TRANSCOM. It entered the war, DESERT SHIELD in particular, with two hands tied behind its back; undermanned, undertrained, underused, under-stressed. TRANSCOM only had several hundred people. We actually fought the war from that old commissary building as TRANSCOM Headquarters. So, that adage, "You don't know about yourself until you're stressed." We didn't know about ourselves at all. As for me, I arrived one week before DESERT SHIELD kicked off. One week before we stood up our Crisis Action Team (CAT) for the first time.

When DESERT SHIELD kicked off on C-day, 7 August 1990, everyone was three weeks behind the timelines in our own war plan. TRANSCOM was crucial to the front-end mobilization and deployments. By all measures, TRANSCOM was further behind. When everything kicked off, we immediately realized we were seriously undermanned in both numbers and expertise. Quantity first. I had to go over to the vice commander of MAC and beg for a handful of people from there to come help us, not trained experts, just people. We were learning our jobs on the fly. We got a few here, a few there. We actually had a reserve unit that had been assigned to TRANSCOM for two and a half years. It was a NR unit, geographically stationed in St. Louis and detailed to TRANSCOM. Naval Reserve (NR) Unit 118 had 92 people in it—wrestling coaches, cabinet makers, lawyers, store clerks, people who worked at Wal-Mart, people who didn't have any training in transportation. To me, they were pure gold because when mobilized, we could man our CAT 24/7. They didn't have any expertise at all, just a repository of local Naval reservists with a mix of everything who were all wonderful people, and a lot of admin. Frankly, in the final estimation, we could not have done TRANSCOM's role in DESERT SHIELD/DESERT STORM without the NR 118. The Navy in the next cycle selected NR 118 as the Reserve Unit of the Year for their contribution to DESERT SHIELD/DESERT STORM.

While manning was critically short, expertise was even shorter, especially JOPES-trained experts right at TRANSCOM, but also at CENTCOM rear and forward, as well as the 35 key onload bases and dozen off-load locations. Nor did we have enough WWMCCS terminals, desktop computers, or deployable laptops. We had but a few desktop computers. We didn't have any laptops. Moreover, exiting transportation IT systems were arcane, legacy, service-specific computer programs that didn't interface.

When DESERT SHIELD kicked off on 7 August, it proved to be quite chaotic. Without designated "traffic cops" and disciplined priorities, people at headquarters and deploying units in the field were inputting to JOPES, asking for pickups of mobilized units in the deployment order. Untrained people were changing pickup priorities. In the first week, we were seeing up to 20 priority changes per day. But, you don't take a large unit that's number 20 priority and make it number one, then, wait a minute, take the 18th priority and make them number one, right behind that. But that's what was happening. Special Operations units were self-deploying partially, then needing TRANSCOM support for the rest. Air Force fighter units were cutting back door deals with tanker units, then demanding immediate airlift without CENTCOM approval. Then multiply that dozens of time each day. The OPLAN TPFDD didn't match reality from the very outset. JOPES allowed people out in the field to go in and just change the priorities on the WWMCCS terminals themselves. CENTCOM forward and CENTCOM rear were wondering what the hell was going on. So, we had poor discipline, lack of training; ultimately, we had chaos.

On 12 August, after all this chaos and gridlock, nobody had any faith in what they were seeing on JOPES screens, and of course nobody had any faith in the TPFDD. MAC was frustrated with dozens of cancellations and re-scheduling's of C-5s and C-141s each day. Of course, after a while, it was just "garbage in, garbage out." Deploying units didn't have any trained people to sort through the chaos. Everybody was pulling their hair out because they wanted to get their people and equipment transported.

By the way, we really didn't have tankers as part of the matrix like we have in global air mobility today. We didn't have any tanker bridge, as you know it now. There was really no concept called "air mobility" back then. It was all airlift moving into bases. We had aircrews, but only 40 percent of our aircrews/pilots were air refueling qualified, and when you looked in the system, you didn't know if you had qualified crew to refuel, even if you had the tankers. And the Air Force held back the KC-10s which were supposed to come to TRANSCOM, so we lost lots of airlift and tankage right there.

So, on 12 August everything seized up. JOPES ceased to function. Gridlock five days in. We reverted to pencil and paper. At this point, CENTCOM stepped in with decisive leadership. CENTCOM let everyone know that only they set the movement priorities. Each day, CENTCOM reaffirmed the priorities for the next 24 hours, then it was 48 hours, then 72, and so on. We reverted to building "Humpty Dumpty" one-step at a time. GEN Schwarzkopf directed his Joint Operations Director (J3) to set the daily priorities. For DESERT SHIELD, it was Maj Gen Dane Starling, who was a CENTCOM Joint Logistics Director (J4) in theater, connected to me as TRANSCOM J3/4, and to Lt Gen Gary Mears, then Joint Staff J4 for GEN Colin Powell who formed the "Logistics Triangle." Through daily, even hourly coordination, we built out a TPFDD for the next 24 hours, then 48, then 72, then 96—eventually getting it all the way out to a week of stable TPFDD. Then, our collective experts inputted it to JOPES for all to see and execute. Of course, Gen Starling was wedded to GEN Schwarzkopf, and I was wedded to TRANSCOM Commander, Gen Johnson. Those were the major senior players at that time.

Simultaneously we spread our JOPES-trained TRANSCOM Masters worldwide to all the key places with deployable devices to input for the customers and also train them at the same time to do it themselves. Once we got JOPES up on its hind legs, we started walking a little bit. We sent them first to CENTCOM forward, then CENTCOM rear in Tampa [Florida], then to key on-load and off-load bases. Before we did all that, we were using the grease pencil methods used in the Berlin Airlift, but using their time-honored successful processes.

So, to get back to Gen Starling and myself. We started with basic things. We went back to basics. First, some key history. Lt Gen William Tunner, who was the father and executer of the Berlin Airlift, was a disciple of Frederick Taylor and his management system. He applied all the Frederick Taylor modalities to the Berlin Airlift [he was also the driving force behind the earlier Hump Airlift in WWII]. Gen Starling and I applied Tunner's lessons learned that were most applicable to DESERT SHIELD. Fortunately, Gen Starling had the confidence of GEN Schwarzkopf as his Chief Logistician. He worked closely with the CENTCOM J3. They stabilized the requirements for each 24-hour period and drove the discipline to keep them stable. Changes were happening all the time, but the next 24 hours was frozen. Then later the next 72 hours were frozen. That was very important to our commercial operators. With our Army, Marine, and Air Force customers, we'd say, "We know that's important, that's moving up, but the air-load plan for the next 24 hours was frozen. You can go in on the 25th hour. We'll put you in there." And we kept building that out, extending the discipline. Then we had to make sure everybody knew about it. JOPES came out of gridlock on the 25th of August. Two weeks inside David Easton's black box, not knowing what's going on in the inside.

While we were sharpening the TPFDD and giving it visibility in JOPES for TRANSCOM's components who were executing the pickups and on-loads, we were rebuilding mutual trust all the while. Getting better, more efficient and effective hour by hour. Continuous improvement as a joint team focused on the DESERT SHIELD mission objectives and timelines. Saddam Hussein helped, he did not invade Saudi Arabia. And he was doomed.

At this point, it was the transportation players having trust in one another and then communicating it via the right kind of systems. We took an Air Force System called Global Decision Support System (GDSS), which eventually grew into the Global Transportation Network (GTN) after the war, and really relied heavily on it to schedule and execute our airlift deployments. So, the things have lasted for 15, 20, 25 years [GDSS, GTN]. We used those systems because they could input into JOPES. Still, turbulence kept things interesting. For example, even the elite and quite deployable 101st Airborne just couldn't get things they needed to get ready to deploy, so they cancelled twenty-eight C-5 on-loads in a three-day period. It gives you a sense of the chaos, which doesn't even count the twenty or so C-141s that they also cancelled. Just think of what that does, it frees up planes, it frees up crews. So, we search for somebody else who is jumping up and down, ready to go. Then CENTCOM must approve it and we redirect that many assets to a new customer. So, this was the stuff that was going on. The 82nd Airborne Division had massive airlift requirements early on, but they lacked a lot of equipment to prepare their cargo and rolling stock for airlift. So, we applied General Tunner's Berlin Airlift rules to the 82d at Fort Bragg, North Carolina. We told the 82d that an airplane would be scheduled to open its doors once an hour on the famous Green Ramp. One C-5 or one C-141 every hour and we were going to do that until the civic leaders from Fayetteville, North Carolina, get on the plane. Then we would know we were done. We did just that. I'm saying we—obviously, then Maj Gen Verne Kondra, the MAC DO, magnificently orchestrated these things. It's not like we at TRANSCOM were doing the missions. We tasked our Transportation Component Commands (TCCs). It was MAC that made it happen, flying those

4,000 C-5 trips, those 9,000 C-141 trips 9,000 miles each way. So, that's the kind of thing we did until we finally got JOPES smoothed out. Changes happened. But only CENTCOM J3 could priority change, no one else. Once it was in the frozen window that was it—weather permitting.

But stuff happens. We discovered that sand gets into the helicopters. Really into everything and that generated the need for spare parts, right now. So, we set up Desert Express, a C-141—sometimes two—every day out of Charleston, North Carolina. Cargo received a certain high-priority code—9AU. I still remember the code. [9AU-project code gives premium distribution handling for direct airlift to the AOR.] We got that item 9AU blessed by the user forward. The shipping unit had to get the items to Charleston on their own either by FedEx or by truck. If it arrived before 1030, then it would get on the 1400 flight, and that flight went direct. The Desert Express crew was air refueling qualified and that plane went all the way into Dhahran [Saudi Arabia]. We also had a European Express later on and even a Sealift Express, believe it or not. The containerized Sealift Express was 23 days, which that was still fast, because of the 13,000-mile distance. The airplanes traveled 9,000 miles. Of course, the Berlin Airlift was only a couple hundred miles and they flew the same few corridor routes. There also weren't any stops or stages in the middle back then. The Berlin Airlift aircraft got in and got out. Air and sea transportation for DESERT SHIELD/DESERT STORM was much more complicated. By sea, it was 11,000 to 13,000 miles to transport requirement and so these are phenomenal distances.

To give you a sense of the size of Operation DESERT SHIELD, in the Berlin Airlift, they flew in food like dehydrated eggs, dehydrated milk, and loose coal to keep the population from starving. In DESERT SHIELD, we were moving tanks, precision guided munitions, lots of Hazardous Material (HAZMAT), weapons, bombs, and all sorts of things. In terms of million ton miles, we did the Berlin Airlift every five weeks. The Berlin Airlift took place over 65 weeks. It's not really a fair comparison, because they were saving the world and altering history. We didn't save the world. We just completed the mission to push a despot out of Kuwait. But if you looked at other measures, our daily tonnage over that distance was 11 times greater than the Berlin Airlift. That induces all kinds of unknowns and variables.

DESERT SHIELD was just an extraordinary thing in the end. No one's breaking their arm patting themselves on the back. When you talk about leadership, I was just the J3/4 of a fledgling, untested joint command. Just a brand new two star dropped into the job as things cooked off. And just a staff guy. But the view from that chair in the TRANSCOM CAT let me see what needed to be done. One has to learn how to lead from the staff. The only way you do that is to do all the things, we've talked about earlier in this interview, listen, check your ego at the door. Construct an effective, matrixed global team of subject matter experts who trust each other. Operate honorably and see the honor in all others. You have to build trust and have a level of situational awareness that the quarterback needs after the ball is hiked, watching the melee unfold in front of you, but make order out of it through team action. These are the types of skills you find in certain staff officers, but not others. So part of leadership is going around and making sure that in every position in the CAT and in your global matrix is competent, trustworthy, knows one's place, doesn't fight the system, and doesn't care who gets the credit. That's a person who understands what it means to buy into the team and takes responsibility for one's actions. The buck stops on their desks. They are tireless, very competent, and have street credibility, with not just with their buddies in the CAT, but also with those they are connected to in CENTCOM forward, at the JCS, and over at MAC, and similar places. If you've got a person that the matrix can't work with, you have to move that person out and put somebody else in. We had such people at the beginning of DESERT SHIELD/DESERT STORM in key senior officer and middle level roles in our headquarters and components. We made the changes quickly, and their replacements shined just as quickly. Most importantly, mission performance improved noticeably—and keep improving. I recommend you do not forget this critical leadership responsibility.

Chemistry matters, a lot. Gen Starling, Gen Mears and I all had superb chemistry. We remained fast friends ever since DESERT SHIELD/DESERT STORM. Contacts matter a lot too. My ACSC '75 classmates are a classic example. While TRANSCOM J3/4 I had to work daily with the Strategic Air Command (SAC)/ DO, Gen Howell Estes, who was running all the tankers, as well as the bomber deployments. And with the TAC DO, Gen Mike Ryan, who was deploying all the fighters. We were all ACSC classmates. Later, as Director of the Joint Staff, toughest job I ever had, Gen Estes was the Joint Staff J3. Mike Ryan was the Assistant to the Chairman—the Mr. Outside to my Mr. Inside role. Years later, when I was TRANSCOM Commander, Gen Estes was Space Command (SPACECOM) Commander, and Gen Ryan was USAF Chief of Staff. Contacts—pure gold—a priceless form of "Professional Capital."

You talk about the importance of relationships and having the right person in the right seat to accomplish that mission. When you look back on DESERT STORM, what would you do differently?

Almost from the outset, all the key players in the deployment parts of DESERT SHIELD [Joint Staff, CENTCOM, our component commands] immediately recognized that TRANSCOM should have had the peacetime authorities before DESERT SHIELD/DESERT STORM. So, they were patient with our fits and starts. This is a positive sign of the global matrixed team pulling together.

So, if you ever get in a situation where you see that the powers in charge are violating the basic principles of "train as you will fight" or "same as peace in war," just to get a compromise, please raise the historical example. If that doesn't work, fall on your sword. Frankly, with different players in key roles, the DESERT SHIELD/DESERT STORM deployments could have imploded and gone the other way. People saved it, and they saved it by using common sense, professionalism, skill, and trust. We had the piece parts, but that is just not enough.

So, that's the one thing we should have done differently. It would have made a massive difference because we would have had an executable TPFDD, trained cadre, and a disciplined process from the outset to handle the inevitable audible we all face at the start of a major military operation or crisis response.

One thing I didn't cover is we did not have something we take for granted today, whether ordering an item from Amazon or deploying to the sandbox, we had no in-transit visibility (ITV). We had no overarching ITV system. The subsystems we had, service systems, did not talk to each other. There was no way to input it. Not so everybody could see it. "Where's my stuff?" became the mantra for this gaping shortfall in our overall IT. JOPES, when working had a little of it. There was a little bit in the Air Force's GDSS, but it was mainly about aircraft tails. It wasn't like we just didn't know what was on the plane. We didn't even know the airplane was coming until it was on final approach to Dhahran. That's the kind of stuff that was going on. Things you take for granted today, like the critical, accurate, real-time info at your fingertips. We had none of it. Or worse, it was there, but it was wrong.

The Army wanted their troops flown into Dhahran—our main aerial port of debarkation (APOD)—just as their shiploads of heavy equipment and containers were being off-loaded at the nearby mega-port of Ad Dammam. Then they could marry up everything efficiently. Ad Dammam, Saudi Arabia had 60 piers and Al Jubail, Saudi Arabia had 20 piers. Those were the two big seaports of debarkation (SPOD) we used.

Dhahran, Saudi Arabia was the big mega-airbase there. But, to TRANSCOM and MAC, Dhahran was like a garden hose at the far end because we had 35 on load bases feeding into Dhahran—so fire hose to garden hose. Match that flow with poor ITV of aircraft tails and onboard pax/cargo and you start to see the potential problems. By the way, we moved 504,000 troops, largely by commercial 747s. But, C-5s could carry 72 troops at a time even with a full cargo load. We made sure we had "back up ITV." A floppy disk always went with the airplane or the ship, so when it arrived anywhere, the data could put into the system en route or at termination and the Army, Marine, or Air force end-user could use it in their system. Then we tried to use all other possible IT systems over and above the floppy disk. This is just the kind of things we did to get the mission accomplished.

Let's talk about the commercial side of the airlift for a few minutes. Let me give you a sense of the passenger movements. About 504,000 [troops] moved largely on commercial airliners. Our partnership with commercial carriers was essential. In the 1950s, we established the Civil Reserve Air Fleet (CRAF). For nearly 40 years, commercial carriers who committed cargo and passenger planes to DOD for use in crisis and war would get government business at prescribed rates in peacetime. CRAF has three stages, but no CRAF stage before DESERT SHIELD/DESERT STORM. We started on 8 August with CRAF volunteers—15 cargo and 15 passenger aircraft. But, on 17 August, we wrote the order to activate CRAF 1, which is activated by the TRANSCOM Commander. That gave us 21 more cargo and 17 more passenger aircraft. Very substantial capability. But, as we surged for DESERT STORM, while having high sustainment demands, we needed more. So, on 17 January 1991, we wrote the order to activate CRAF II, which was activated by the SECDEF, who at the time was Dick Cheney. That raised our commercial CRAF to 76 pax and 40 cargo planes. The President activates CRAF III. We actually wrote the order for CRAF III later because we needed even more cargo B-747s. We actually got some additional volunteers, sort of "CRAF II-Plus." You actually tank refuel, as you well know. So, those numbers are meaningless to people.

Together, our military and commercial airlifters moved the equivalent of Atlanta, Georgia nearly 10,000 miles to the desert, then we moved a lot of it back after cleaning it, by the way. That's all the people, all their possessions, all their vehicles, personal and commercial, and all their stuff, their inventory, all their food moved by air. So next time you're in traffic in downtown Atlanta, take a look around. We moved that, in a period of seven months. So one Berlin Airlift every five weeks, and just by air, we moved Atlanta.

By the way, normally when we're doing official operational planning, we plan to move 10 percent of everything by air, and 90 percent by surface. In DESERT SHIELD/DESERT STORM, we moved 19 percent by air and 81 percent by surface.

Why? To catch up. We like to say, "Airlift makes up for the sins and errors of others." To repeat, the CENTCOM OPLAN (with no TPFDD) assumed 20 days notice and 10 days to move into position, so 30 days. But by the time the Pentagon issued the deployment order for us, we were 21 days behind. And yet, we almost made closure at 15 January. That was closure of forces far exceeding the original DS deployment order. The final force closure unit deployments just kept growing well into January 1991. The total movement requirement nearly doubled.

The major reason for the growth in the total requirement was the decision to move the 2d Armored Division from Europe. It was called "The Surge." And it occurred in November to January, on top of the already-existed sustainment spike and the holiday mail spike. It created a "super spike" transportation-wise, and it was all very sealift-intensive as well. To complicate matters, we experienced the worst winter weather in 20 years. If the Cold War had not stopped the year prior, 2d Armor would not be available. They had no mobility mission, which meant they had no mobility training (including JOPES training), no mobility equipment, and no mobility manning. So 2d Armor was very heavy. The Surge also placed extreme emphasis on the previously-established force closure Date of 15 January, so all would be in place to execute DESERT STORM operations as planned. A lot of 2d Armor moved down to the port of Bremerhaven [Germany] by river barge because the roads were too treacherous to bring tanks on with black ice all over the place. All kinds of interesting things took place in those winter months.

Can we transition to surface transportation? Can you give us some insights into that?

Now, regarding sealift. On C-Day of DESERT SHIELD, 7 August, sealift kicked off simultaneously with our airlift ops. A couple interesting things on the surface side. Everything I said about the first time we activated the CRAF, well, same on the surface side—lots of "firsts" activation-wise. Organically, TRANSCOM's Navy component is the MSC. MSC did two things immediately on C-Day. First, it alerted two of three maritime preposition squadrons for movement to the AOR ports. Second, MSC activated its organic ships on hand for rapid deployments. These are seven Fast Sealift Ships (FSSs). These ships are huge and fast, designed to move heavy Army divisions from designated ports very quickly, days and weeks. These are the C-5s of our sealift fleet, with 220,000 square feet of deck loading space plus other dry cargo and container load capacity. FSSs are capable of over 30 knots speed. All else sealift-wise exists in ready reserve status, or is contracted for with US-flag and foreign flag commercial shipping, bulk dry cargo, container, or POL tanker.

MSC has control over the Ready Reserve Force (RRF), 92 government-owned military useful older ships. But, the RRF had never been activated en masse. Just a few ships have been activated here and there for annual exercises. The RRF's 92 ships in turn were maintained on both coasts by Department of Transportation's (DOT) Maritime Administration (MARAD). So, the RRF were mostly older ships owned by the government, but crewed by civilian crew members of the Seafarers' Union. These ships were largely loaded by the Longshoreman's Association. Like CRAF, the RRF is activated, called "broken out" incrementally on different levels, break out times of 5, 10, and 30 days. Each level has different response times. And there are different kinds of ships. The favorites were activated first, 24 Roll-On/Roll-Off (RO/RO) vessels, each averaging 100,000 square feet of loading deck space which everybody loves for loading tanks, other armor, heavy trucks, Humvees, and other rolling stock. Our RO/ROs were 1 to 20 years old. The rest of the RRF ships are "break-bulks," which are primarily 10 to 30 years old, and carry dry cargo, nonrolling stock, and ammo.

So the RRF was critical to the early movement of our key Army and Marine units. But like much else, the adage, "no one knows about one's self until one is truly stressed," applies here too—in spades. The RRF was chronically underfunded. And the crews were older reserve types who had other civilian jobs. So, pulling it all together on critical timelines was a major story in itself worthy of a novel.

The first thing we wanted was the 24 RO/RO ships out of there right away, to get them to move it to Jacksonville, Florida and get it to Savannah, Georgia, because the 24th Infantry Division is coming out of Fort Stewart, Georgia. These RO/ROs were all going to be there in a five day breakout, well, they couldn't even find the mariners. Fortunately for us, there was the Seafarers' International Union, a bunch of patriots—TRANSCOM is bonded at the hip with them forever—and the International Longshoremen's Association, who made these activations and critical early on-loads work. No other command does this and still doesn't to this today. The bond is forever, because these patriots—they stopped working as assistant managers at McDonald's—

and they came, got a little top-off training, and they went on their ship. These ships, instead of breaking out in 5 days, they broke out in 10 days, but they got force closure of 15 January and earlier.

TRANSCOM's Army Component was the MTMC. On C-Day, MTMC sent port teams all over the US, Europe, and the Persian Gulf to manage all the loading, off-loading, and everything else. These small teams, all reservists, are priceless. They had the expertise to create the floppy disks to go on the ships. They are the heartbeat of America's sealift bridge.

Before the global sealift, there is "Fort-To-Port." This is an epic operation in itself. MTMC also had 1,600 heavy railcars that went out to the forts that were loading tanks, other massive heavy equipment. The rest of the loads, we used over 16,000 railcars to get the Army and Marines to the Continental US (CONUS) and European coastal ports. Also, MTMC used 54,000 trucks in addition to the 16,000 rail cars. In many cases, we also had to get the people to travel with their equipment and maintain it en route. And during this time, America's number one bus company decided to go on strike. Other bus companies pulled together and made it work.

How about safety? Out of the 54,000 trucks, we had one accident. How good is that? How about being blessed by sheer luck? We had 16,000 railcars, and we only had three derailments, which were on re-deployment, all caused by deteriorated rail spurs that had been weakened on the outbound deployment. Here again, we had not stressed ourselves for decades. Everyone really pulled together; a true military-civilian partnership that worked. In the culture of our deployment, no one ever shouted at each other or got emotional on the deployment. But, on the re-deployment, people did get emotional because they wanted to get home. People realized they could get emotional without penalty—human nature—very predictable.

Some other surface transportation observations. The Army was not container friendly—hard to believe today, I know. Again, largely unstressed for decades. We had an Army that logistically was still in Vietnam. Everything was unit break-bulk to get it in there, as opposed to containerized. The Army had built in plenty of time to get out to the battlefield. They kept their rapid mobility units lighter than normal. These units didn't envision heavy armor desert warfare. They didn't have a lot of heavy equipment, because tanks and other heavy equipment weren't really used in Vietnam. For DESERT SHIELD/DESERT STORM, the Army requested that all their heavy stuff that was going forward to come into one place (Ad Dammam). They wanted to marry up everything with their troops arriving at airbases nearby. Not ever done before on this massive scale. Of course, TRANSCOM, MTMC, and MSC were trying to maximize ship loads. This was at odds with the Army's requirement for maintaining unit integrity, which might actually add up to a 9.2 ship's worth. Well, we wanted to load out that tenth ship to reduce waiting backlog. Same thing with containers. The Army didn't like containers, because they didn't train to put their heavy equipment in a container. This was the first container war. We had lots of container ship capacity available commercial and it really speeded up on-load/off-load times. So together, we had to build out all of the container culture and all of the databases from the commercial sector. The Army didn't own any containers. TRANSCOM had access to 17,000 containers. We had to have a mind meld. Another example: The Army didn't like to send ammo the old fashion way, piecemeal on breakbulk ships that required long load times and large amounts of wood shoring. It took 7 9 days to load out that way versus 72 hours by container. Thus, we had another mind meld. You begin to the value of creating a TRANSCOM with wartime authorities, but also why peacetime authorities are needed too. We would have loved to have used far more containers. We would have gotten 20 percent better closure by using containers. As an aside, this applies in airlift too. Had we had the C-17 [Globemaster], we could have gotten 20 percent better efficiency than using the retired C-141 fleet. These efficiency opportunities were all over the place. When we got to exploit them it made a difference. And when we didn't, well...

CENTCOM, God bless them, would not give us an in-theater aircrew stage. I mean, look at all the bases and ramps they had, but they would not approve an in-theater stage. Again, I hate to use the term 20 percent, but we would have had 20 25 percent more strategic airlift efficiency if we had had an in-theater stage, but we never got it. Cost nothing, but would have been the equivalent of having C-17s instead of C-141s. If TRANSCOM had control of the tanker fleet like today, including the KC-10s, we could have squeezed yet another 20 percent efficiency out of DESERT SHIELD too. On the sealift side, after the war we procured 24 Large Medium Speed RO/ROs, each with double the capacity of the ones we used in DESERT SHIELD/DESERT STORM. And, the Army had a transformation epiphany regarding containerization and all the necessary accouterments and training. Then there is the IT improvement which is correctly described as three generations beyond what we had in DESERT SHIELD/DESERT STORM. We are light years beyond what we had almost 27 years ago.

And so, when DESERT STORM ended, when we declared victory and the cessation of hostilities, we had a ship on the ocean every 50 miles en route to the Persian Gulf. Do the math. We turned around over 100 ships carrying ammo, containers, resupply, that we had already put in motion. They just made a U-turn and came on home. In OIF, when the US pulled out of Iraq, J4 crunched the numbers on how much it would cost to bring equipment back versus just leaving it there. It was more cost effective to leave it, and millions of dollars of equipment was left behind. As in all wars ironically. The waste is shameful when you think about war. Another reason to avoid it whenever possible. You know, the American public never gets a sense of actually how much we leave. You made me think of a couple of things. Pallets. Pallets were a "roach motel." Pallets went in, they didn't come out. We didn't have any ITV. The Army just told us to put the stuff on a pallet and put the pallet on a truck and off it went into the desert on pallets. One pallet at a time. During the surge in late 1990, we almost ran out of pallets. Next time we'd see a pallet, it was part of the fortification they erected in the desert. Same thing with containers. Our commercial operators came to us and told us, "In two more days, we're going to be out of containers." "What?" So, we had to go out and get the containers from the desert and bring them back. We had to develop all the process to maintain control over the containers, pallets, and pallet netting and tie downs too. At the same time we're doing that, we're buying pallets, nets, and tie downs. A pallet is only good if it has a net. Boy, those nets were good for a lot of things in theater, so the nets disappeared, tie-downs too. So we had to get all the processes up to maintain control of this material handling equipment (MHE) that we needed badly all the time, but we take for granted, completely unknown until you are stressed.

Talking about MHE, planes are one thing, but what really makes air mobility work is the Airlift Control Element (ALCE). ALCEs are the jewels of throughput. Without throughput, there is no air mobility. You have to have loaders, receptors, trained people to off-load it, and specialized vehicles and other equipment to load and off-load rapidly. During DESERT SHIELD/DESERT STORM, we didn't have reliable off-loaders. The machines, called K-loaders, were Vietnam-era and unreliable. Mean Time Between Failure (MTBF) was often a few hours for our K-loaders. Throughput at the receiving end became our largest limiting factor. After DESERT SHIELD/DESERT STORM, we bought an entirely new fleet of ultra-reliable K-loaders, called Tunners and Halvorsens, named after our two iconic Berlin Airlift heroes. We got all of our planes going and all the TPFDDs going. This stuff is working great! "Wait a minute—we need two more 40 K-loaders to get the stuff off." We just couldn't get it off. Ground refueling was another challenge. We needed more fuel trucks on the ramps. Some planes waited for fuel 8 or 10 hours. These things you don't think about when we were getting so efficient. We were delivering it all so fast, that success in one area created problems in other areas. More lessons learned, but revealed only when one is fully stressed.

A master metric. For every troop we moved into theater, you moved one ton by air and a total of 20 tons by surface. This is true even today. These are planning concerns, because if you build a TPFDD and you don't see those kind of macro relationships, that means something isn't right or someone has achieved some sort of efficiency, but you'd like to know what it is, because you want to make sure they didn't assume it away while building the OPLAN. It might not be there in real life. These are the lessons you learn from young majors and NCOs. It's why you have to think, listen, and have situational awareness. Listen, be humble, and see honor in your teammates.

We had the privilege to hear from General Darren McDew, who is a close friend of yours as the current TRANSCOM commander, and had the opportunity to ask him what would be the one question he would ask you if given the chance. His question was, "How would you fix the cyber challenges that we are dealing with today?"

I'm generationally separated, even from Gen McDew. He has massive and impressive practical experience in air mobility and defense transportation over three decades and thought out our Nation's two longest wars. It is kind of him to ask for my input. But I am a mere relic now. I tended to study the ones around me who were pure gold. And I studied the masters who had achieved and who had published. I was far more interested in studying Frederick Taylor right down to the nitty gritty. How General William H. Tunner conducted the Berlin Airlift in third level detail. First level detail won't cut it. I skipped over trying to learn from my immediate senior Statesmen because most of my predecessors had no wartime experience in the job, at least no experience in fully stressed airlift or air mobility. From Vietnam all the way through, until we did DESERT SHIELD and DESERT STORM, they might have been in the job for three or four years, but they didn't have any real experience in what I was experiencing. Just Operational Readiness Inspections (ORI) and exercises. And, I knew those events proved deceptively hollow. Also, I felt I had enough situational awareness to know that because I was underneath them and knew what was done at that time, so I didn't actually seek them out.

Regarding General McDew's question, we didn't have any cyber challenges in DESERT SHIELD/DESERT STORM. Our enemy was not competent in this area. That was one of the benefits. Looking back, cyber threats would not have made it on to the lessons learned list.

We did have a real weapons of mass destruction (WMD) challenge. It was real. And it had a rapid onset. SCUD missiles with chemical warheads that were fired at Israel. They fired the first missile and Israelis said they needed Patriots, and they didn't have any. So, we had a requirement to get three Patriot units to Israel within 48 hours from the deployment order, in place and operational, with extra missiles. We sourced it from EUCOM. We beat the timeline. We actually got it there. Remember, there was lots of talk of chemical warfare with Saddam Hussein—it was a serious chemical warfare threat. Throughout the AOR, most of the people for months on the ground walked around with chemical gear on all the time. Flying missions were conducted in chemical gear. We had to do chemical warfare training with our commercial planes. Commercial aircrews had to be trained in chemical and biological. We actually had "Balkers," those commercial providers who refused to enter the Persian Gulf after the SCUD attacks on Israel. We not only used US commercial providers on sealift and air, we also used foreign flag carriers on both of those, particularly at sea. Thirty-four nations provided ships carrying our cargo. Well, a few would get to the Straits of Hormuz, and they'd be picking up all this stuff too about the chemical threat on the radio and say they weren't going any further. We worked off that minor anomaly.

We had pirates. And, do you know who the pirates ended up being? They were the foreign workers in Kuwait who lost their livelihood when the Iraqis came in and took over. They left everything back in Kuwait; they had nothing so they were just pirating feeder container vessels transiting the Straights of Hormuz. And, amazingly we only had one case of pilferage in the entire operation. Terrorism? We had bomb threats in Europe and even two actual bomb attacks against MTMC Headquarters in Bremerhaven [Germany]. These things never make any of the history books now, but they actually happened. Lots of bomb threats, more than you could ever count. They would slow operations at the ports. So, port security was tightened very early on at on-load ports and especially around our main ammo ports. An ammo loading port doesn't have to be tight, it has to be airtight. You get a bomb threat against something like that, it is serious business. So, we had those kind of things and got through them.

I have heard from three different generals now that the balance between family and career is constantly a question that comes up. So, with 34 years of active duty, how did you balance your family and work life?

Well, first of all, I must say that we're still engaged in two of the longest wars in history, and it is the both of you and all who serve today that have lived at such an extraordinary heightened pace. My hat just goes off to you. I truly admire all of you. You've had to live with this at such a sustained high level for so long.

I was very lucky that my wife, Kay, chose me in 1966, rather than someone less tolerant. She always knew that I was going to give it a hundred percent and she was going to be very supportive of me. Her father was a doctor in WWII, and he was gone overseas in Europe for three and a half years. So, Kay knew what might lie ahead when she chose me. He was the town doctor in Opp, Alabama for 42 years, and this was back when doctors made house calls. He was my role model in life. So, she grew up in a family of service and sacrifice. She knew all that entailed. She has always wanted for me to be successful in whatever I wanted to do and she has had an unwavering sense of national service and this is how she fulfilled that intent. While at ACSC, our daughter Karin came into our lives. During the Pentagon years, I went in at 0600 and came home 1900, often later, plus weekend work there too quite often. So, this was our normal life. My time with Kay and Karin was evenings and bedtime stories plus all the weekends we could cobble together. Pretty average stuff really and we loved it. Six straight years in the Pentagon, 3.5 as the Defense Suppression Desk Officer in Air Staff Plans, known as the Fighter Mafia back then, and 2.5 years in the Chief's Staff Group as the Theater War Guy; led to two more early promotions and a National War College [Washington, D.C.] slot where I spent evenings writing a book on military reform. So that's seven years in downtown D.C. Tack on two more years at the 89th at Andrews as wing DO and wing vice, and you can see why our daughter felt she grew up in Washington, yet we moved three times. Her favorite weekend haunt was the Smithsonian.

Our daughter Karin really grew up making the real sacrifices. After age seven, she was the one who was left with the babysitter when I started to be a colonel and general. This was still the old school time, when Kay and I had official obligations four or more nights per week, Karin grew up on TV dinners from the microwave. After Andrews, we spent over three years in Dover, Delaware where Karin spent most evenings home alone. And, after I got promoted to brigadier, we moved to Randolph AFB [Universal City, Texas] to be the Vice Commander of the Personnel Center, and then I became the Plans and Programs Officer of Air Education and Training Command (AETC). While we were there, we picked the T-6 [Texan Trainer], the T-1 [Jayhawk], and published the Trainer Aircraft Master Plan. All of these things have legs right to this day. We even planned the T-38 replacement as the T-X in the master plan. Finally getting to it now. We left there for Scott AFB and went right into DESERT SHIELD/DESERT STORM.

As we finished the redeployment and stood down the TRANCOM CAT, I left immediately for the Pentagon as the Headquarters USAF Director of Operations. Five months later, as USAF elected to consolidated SAC, TAC, and MAC into Air Combat Command (ACC) and Air Mobility Command (AMC), I was reassigned back to Scott as the Provisional Commander standing up AMC. Kay and Karin stayed back in Virginia so Karin could finish her third high school in four year as Salutatorian of her McLean (Virginia) High School class. Her third high school in four years. Then she entered Rice University in Houston (Texas), "With a lifetime of experience," as she likes to say. Today, she and her husband live in Austin (Texas) where they work for cloud companies and enjoy their first child. So, this was our normal. It's just in the make-up of our family. They were there to support me and we just tried to make it work, just like you do. I don't think it's any magic formula. You just do it. It's that simple.

I went on to four more assignments after standing up Air Mobility Battlelab in 1992. I was AMC's first vice commander, then the 20th commander of the venerable 15th Air Force. A two year tour as Director of the Joint Staff in the Pentagon followed—when they tell you your office comes with an attached sleeping facility, run the other way. Finally, I was paroled to my final assignment at TRANSCOM's fifth commander, and fourth Commander of Air Mobility Command.

I would like to leave you with a few quotes to take with you as you continue your career:

- No one knows about oneself until truly stressed.
- No organization knows about itself until it too is truly stressed.
- "Never make a decision before it is time."A favorite quote by Colin Powell.
- Human nature in an organizational setting is a highly predictable set of near physical laws, predictable to 95 percent certainty—so stay humble and listen carefully.

■ Interviewed by Maj Heather Demis, US Air Force, and Ch (Maj) Christian Williams, US Air Force.

Feb 1965, Pilot training.

Oct 1966, F-4, Elgin AFB.

Apr 1967, F-4 with backseater, Lt Burke Richardson.

May 1968, Killing time on Quick Reaction Alert Da Nang, AB.

Kross arrived at the Pentagon as a newly promoted major and left as a colonel select.

1984-1987, Flying the C-5AB.

Sep 1986, Picking up the first C-5B, named "The City of Dover."

Taking command of the 15th Air Force.

WINGED ASSAULT
U.S.
U.S.

Rhonda L. Cornum

FLIGHT SURGEON | DESERT STORM PRISONER OF WAR

Brigadier General (ret) Rhonda L. Cornum was born in Dayton, Ohio. At a young age, her family moved to New York, but she would spend large portions of her childhood traveling between the two States visiting her grandparents. While staying with her grandparents in Ohio, she learned how to work hard and how to handle weapons. From her grandfather, a former Marine, she also learned the importance of living a life of honor. He was not the only strong influence in her life as a child, however. The women in her family taught her to be strong and that a woman did not have to be confined by social norms. Her great grandmother was a pilot and was acquainted with Amelia Earhart. Her grandmother showed her that a woman could be a mother, but could also work and have a life beyond the home. As a teenager, she was a headstrong tomboy who had no time to worry about what others thought. She was an outstanding student who graduated just after her junior year of high school. She was then accepted to Wilmington College on a full academic scholarship. Her love of chemistry in high school paved the way for a degree in microbiology. During a lecture she was presenting on amino acid metabolism, she was approached by an Army lieutenant colonel. As luck would have it, the Army was looking for a person to conduct just such research in amino acid metabolism. After accepting an invitation to visit the Army research center in San Francisco, California, Cornum was so impressed she decided to join the Army. In the spring of 1978, she raised her right hand and joined the Army as a medical researcher. She completed medical school at the Uniformed Services University [Bethesda, Maryland] in 1986, and was quickly drawn to the combat field and aerospace medicine arenas. She earned her Expert Field Medical Badge, a feat not achieved by many. She was also selected as a candidate for the NASA astronaut candidate program. Her love of flight grew as she completed airborne, air assault, and flight surgeon training. Another highlight during her time in medical school was meeting her now husband, Brigadier General (ret) Kory Cornum. Throughout her life, Gen Cornum has excelled at everything she has put her mind to, she accepts nothing less. Some of her medical aviation research led to the enhanced use of helmet mounted displays in advanced attack helicopters. During an interview for her book, She Went to War, *she remembered almost missing the call that would change her life forever. While out running errands, and before cell phones, Army Lieutenant Colonel Bill Bryan was feverishly trying to get a hold of, at the time, Major Cornum, to ask if she was willing to deploy to the Persian Gulf. The answer, of course, was, "Yes." On the morning of 27 February 1991, she embarked on a combat search and rescue mission to recover an Air Force pilot, Captain Bill Andrews, who had been shot down. Tragically, the UH-60 Black Hawk helicopter that she was on crashed after being hit by Iraqi anti-aircraft fire. One of only three survivors from the eight-member crew, Major Cornum was captured and taken prisoner by the Iraqi Republican Guard. She suffered two broken arms, a severely damaged leg, as well as gunshot and shrapnel wounds to her shoulder and head. Minutes after the crash, she and the other survivors were taken as POWs. She was repatriated on 6 March 1991 as one of only two women POWs in the Gulf War. Although US law prohibited women from serving in combat roles, her experiences and open dialogue as a POW helped pave the way for continued Congressional expansion of military women in combat. She and her husband share a love of flying and built an experimental aircraft by hand. As a major, Gen Cornum became the first medical officer to graduate from ACSC in 1991-92. She continued medical training and research in the field of urology, commanded the 28th Combat Support Hospital at Fort Bragg, North Carolina, and deployed as the Medical Task Force commander to Bosnia. She was also the first female commander of Landstuhl Regional Medical Center, Germany, leading medical treatment for over 26,000 injured veterans of Iraq and Afghanistan. Gen Cornum's career culminated in founding and leading the Comprehensive Soldier Fitness Center. The Center develops psychological strengthening and resilience training to aid military members in surviving difficult, even life-threatening, situations. Awarded the Distinguished Service Medal, Legion of Merit, Distinguished Flying Cross, Bronze Star, Purple Heart, Air Medal, POW Medal and others for her service, retired Gen Cornum spends most of her time on her family's 700-acre family farm in Kentucky. Gen Cornum has had experiences and challenges during her life and career that most others have not had to face. She provides a unique perspective on what it means to be a successful leader.*

EARLY CAREER

Why did you choose to go into the Army and why did you choose to go into medicine?

The truth is I joined the Army because I wanted to do full time research and I didn't want to teach after I got my PhD and the Army offered the opportunity to do that. They said, "You know we've got a great lab, we need the kind of person who does your kind of research. You'll have your own analyzers, you'll live in San Francisco." That's when Letterman [Army Hospital] was open. The guy said, "To me the only catch is you have to join the Army." Well, I'd never met anybody in the Army so I didn't have any prejudices about that one way or the other and all my other jobs were either assistant professorships or post doctorates. Or, there was one at a nutrition lab, or where I would have been the youngest person, the boss, the only girl, or the only non-Mormon and that did not seem like a great recipe for success. So I decided, "I'll join the Army," medicine came four years later.

So, you had your PhD?

I had my PhD when I joined. I was an unobligated volunteer and got four years of credit toward rank, so I came in as almost a captain. There was no question what I was going to do; I just had to find out where I was going to do it. Then when I got in, I really liked it. I mean it took like a week and I knew I really liked it. So here I am. [Laughs]

How did that translate into medicine then?

That translated into medicine because I went to the Letterman Research Institute and I did research and I wrote grants and I got money and published papers and all that stuff. And I looked around and I said, "I'm really a little irritated, that I'm doing twice as much work and writing twice as many papers and getting paid half as much as the physicians." So, since I don't like to be frustrated and irritated, I could either go somewhere else, where that won't be in my face every day or I could go to medical school. Coupled with the fact that I had done none of the things that a medical services corps officer should have done to get promoted to major. I hadn't led a platoon anywhere. I hadn't been a staff officer on any companies. I had done none of the things that were going to get me promoted. So, since I liked the Army, I said, "I have to do something different than be a medical service corps officer and researcher or I'm going to have to get out." So I decided, "I could either go to flight school, or I could go to med school." Those were two things that looked interesting and upwardly mobile, but at that time 1980-1981, women could fly nothing except, as we called them at Fort Rucker [Alabama], slugs. I couldn't fly [AH-1] Cobra's, so I said, "I do not want to join a profession where I start off being discriminated against," which tipped me to med school. I had no plans to do anything other than to go to med school and get back out and do research, but you know, things change. So that's the story. And it turned out that once I got there I was good at it, and I liked it, which is certainly better than not liking it and being bad at it. [Laughs] I basically believe you are going to be good at whatever you try hard at. If you don't pigeon hole yourself into something and think you can't change. I teach psychological fitness now for a living. The worst word in the world is can't, because that's self-fulfilling. If you ever say you can't do something, you're absolutely right. That's a decision.

What influence did childhood/family have on your decision to join the Army and medicine?

None about medicine. I'd say influence of my family, the first was . . . [turns around and points to black and white picture on the wall] I realized the other day that's the only picture of a family person, in fact it's the only picture of a human I have in my house. It is my grandfather who was a Marine at Iwo Jima and he was the one who told me, "Rhonda, there's worse things than dying, there's living with dishonor." He had sort of instilled that in me. So there I was, I had the opportunity to do something honorable for a living, and that seemed like a good thing. In terms of other influences, I would say that it was probably my dad who said to me, when I was very young, "You can do anything," but then he added, "You can't do everything, so learning to prioritize is invaluable!" That's pretty profound, isn't it? It's one of those things that I talk about whenever I'm asked to speak at a graduation or similar event.

GULF WAR

How would you describe your involvement in the Gulf War?

I was asked. In fact, I was returning from a TDY at Fort Hood [Texas], where we were gearing up guard units to be [AH-64] Apache units when Iraq invaded Kuwait. I was in the Atlanta [Georgia] airport watching this unfold, and I was with an Israeli officer who was commander of their first Apache unit. He looked at me and said, "What do you think?" I said, "I think I better get back!" He said the same thing. So I got back, and by the time I got home, Kory said, "Rhonda, Colonel Bryan is trying to get ahold of you on the phone." He was the commander of the Apache unit that I actually went with. I was actually employed by the research lab there at the time and he asked if I would go with his unit. It's amazing how fast orders can be generated when there's some urgency, by the next morning I was a member of the 229th and we left that week. We were like one of the first ones in. We got there before the gas did, we were just hoping they were intimidated. We were in the same boat as the A-10s, we landed but couldn't take off. We wanted to look scary but we weren't. [Laughs]

How did the Gulf War experience shape or change you?

Well, I would say it was, even though the term didn't exist, my opportunity to demonstrate post-traumatic growth. That experience made me better, certainly a better patient and a better doctor, because I had never been a patient with a significant problem or had a family member who I was worried about due to a significant problem. So, I was much more apathetic before. It certainly made me a much better leader, if the worst thing that ever happens to you is someone yells at you well that can be kind of scary. If that's not even close to the worst thing that's ever happened to you, well, then you can go off and do the right thing because it's the right thing to do and you don't really care if someone is going to yell at you. Much more appreciative of family, when you come really close to losing them you do appreciate them more. I guess maybe, confidence. Confidence and personal strength. You know you always think you can do anything, like my dad said, okay, I know it's true now. [Laughs] It's not that you don't have respect, and you still do risk analysis, but fear is kind of gone. I wasn't ever really afraid before but pretty much not at all after that. I had a surgeon one time tell me when I was a medical student, this is another one of those things like my dad said. This was Colonel Dan Rosenthal. I was a fourth year medical student, and I'm presenting a case about a guy that has an abdominal mass. I wanted to do a biopsy of the mass and I'm describing the procedure. Then I said, "I would be afraid that" and I was going to list off all of the complications. Dr. Rosenthal said, "Ohh Dr. Cornum, you're a general surgeon now, you might be wrong, but you are never afraid!" [Laughs] He was right, it's one of the things we learn—negative emotions don't help you, ever! Unless they lead you to purposeful action to do something different, but fear and being tachycardic and hyperventilating never helps you, whether it's putting lead downrange or operating between the vena cava and the aorta, shaking hands will not help, so don't do that! I don't think I used fear as a description of my emotions since that day. I will maybe say it was an anxiety-provoking thing, but that's a lot less emotionally charged than fear.

LEADERSHIP

What is one word you would use to describe the most important aspect of leadership?

Honesty. When I went to the General Officer charm school, I got there on time, not early, so I'm the last person at this big table of 49 little baby generals and they start going around the room. Who are you, what's your next assignment and what else do you do? Most people said, "I'm General Schmedlap and I'm going to be the deputy commander of wherever and I don't do anything except go to work." Most of them said that. A few said, "I play golf badly." A few said, "My hobby is my family," which I thought was really sad. Then they come to the 49th baby general and I said, "I'm Rhonda Cornum, and I'm going to be the director of Comprehensive Soldier Fitness, and I've got to tell you that I show world class dogs, everywhere." I said, "I took best of breed at the world show in Poland in 2006" and there was silence around the table. Afterward, we went on these trips and every one of those guys came up to me and asked if I really did that, then asked, how I could do that? You know, work will expand to fill the time that you give it.

How sad is it that our military leaders feel that way?

I don't think it's just leaders.

No, but it's expected of everyone else, we only give lip service to family time.

That's why I said honesty was the most important aspect of leadership, I meant it. When I said, "I want you guys to have another life." I meant it. I went home at 1530. I would say, "I came in at 0530. I'm going to go home. You're welcome to stay, but don't expect extra credit for staying, you won't get any. It just means you don't have a life." It's all about honesty. I remember when [Hurricane] Katrina happened, and I was sitting at the operations officer's [G-3] office at the Pentagon. The G-3 wanted everyone in the office every six hours for a meeting, which meant that if you lived anyplace other than next door, you never went home. Finally, it's Saturday and I said, "Sir, what are you going to ask me to do?" He said, "Well, nothing." So I asked, "Well, then why am I coming in?" He didn't really have an answer, so I said, "Well, if you need me for something call me." When I was commander of the Combat Army Support Hospital (CSH) and made a particular decision. My boss said to me, "Rhonda, you realize that any other colonel would have done something different." I said, "Yes sir, I do realize that. Like my mother used to say, I'm not their mother so I don't care what they're doing and besides, what all of those other colonels have done hasn't worked, so I'm going to do something different."

What has been the most memorable leadership challenge you have faced in your career?

That's a hard question because there's been a lot of cool leadership things that didn't seem so challenging to me. The things I remember being difficult, which may not be the same thing as challenging, may have actually happened after I left. Here is the problem, I may not have been a great leader, but I was a well-loved leader, and so when you leave command and you hand off the flag to somebody else, you can't take responsibility for those people anymore. No matter how many times they call from downrange and tell you how horrible things are and how they miss you, and can't you do something and you have to say, "No, I can't. I'm sorry." That is horrible, I had taken responsibility for people's lives and now couldn't do anything. That was for me personally, the most difficult. To somehow tell them I still cared, but I couldn't fix it.

What was the most rewarding leadership aspect?

My most rewarding time as a leader was at Landstuhl Regional Medical Center, being a surgeon when we were the number one, busiest trauma hospital in the world. Luckily, we had a surgeon in charge and we had my husband, who was the commander of all the Air Force people there, so we had the most jointness, and the best ability to get stuff done, ever! That was probably the most rewarding thing for me.

What was your most rewarding leadership moment in the deployed setting?

Probably to be trusted by individuals who would talk to me. When we go to war, we take a wellscreened population, particularly aviators, they're glowingly healthy. What they have though is the same life problems that any other young male or female far away from home in a dangerous and ambiguous environment has. They have personal problems, or they have uesolved issues for a long time. Having people talk to you and trust you is very rewarding. Several people said they had never talked to anyone about their problems, but they thought, "You know, I'm going to war and I might be killed, and I don't want to be killed if nobody knows this." I just thought that was an amazing thing, to be trusted that much!

What was the least rewarding leadership aspect when deployed or during your career and how did you overcome it? Anything that maybe made you question being there?

Nothing made me question being there when deployed during DESERT STORM. It's one of the few times where I think we truly wore the white instead of the gray hats. In DESERT STORM, there truly was an aggressor, that we stopped, made him go home, then we went home, and things returned to normal. Maybe normal wasn't great, but we at least went there. I've never had more than about 15 minutes where I've questioned what I was doing or what was going on. The decision comes down to whether I was going to have to unmake a bad decision someone made, or I was going to have to get out because I couldn't support their decision. Luckily, I was always able to find a way to extract myself from being forced to help promulgate a bad decision. That goes back to honesty. I think you have to be honest. If you honestly believe what someone else is doing, is not illegal, but it's bad for soldiers, bad for the Army, and bad for the Nation you're still obligated to do it. One of the other things my dad told me was, "Rhonda, the fact that you're good at something, does not obligate you to do it." When I was acting deputy commander of a facility and they wanted me to help sell a bad policy to the staff, I said, "Sir, that's a bad policy and I'm not going to try to sell that to anyone, it's only going to save money on a temporary basis; that's not a good reason to demoralize all of the active duty positions you have on this base." Luckily, I got to move to the research facility at Fort Rucker after that. I went home and told Kory, "I'm either going to get a bottom tier officer evaluation report (OER) because of my lack of loyalty or something else." I was amazed. It became clear that the commander knew it was bad too, because he gave me an OER that said, "Towering Integrity." I thought, "Well, if he knew it was bad why was he supporting it?" But, of course, I didn't say it, I just said, "Thank you." [Laughs]. It made me wonder at the time though, "If you know what you're doing is wrong, why don't you just not do it? What's to be afraid of, you might have to get a job on the outside and make twice as much money?"

What leadership lesson would you like to pass along?

I think the leadership thing that made me successful was trust the people to know what they know how to do and save yourself for problems that they can't do. I remember saying to a boss who was going to come with me to do something, "Sir, this will go no faster if two of us go. So if you want to do it, I'll go do something else." If you don't trust the people you have working with you, then find different people. I think the people who don't trust people working for them, don't trust themselves either. Trust them to do their jobs and go do something they can't fix. Go get the resources that they need to do the job you have asked them to do. That's the key, you are the resource maven.

What would you like your legacy to be?

What I do now is similar to what I did in the Army at the end, and it affects large numbers of people. It is probably the most important thing we do, but I don't really care if anyone remembers me for anything. I guess I'd like to be remembered for demonstrating that leadership and courage and vision have nothing to do with gender, positively or negatively.

Who were the most influential leaders in your personal and military lives and why?

In my personal life, I'm going to have to list people I never thought about as leaders but they are the most influential. I'm going to have to say Pete Ziccardi, who was my high school chemistry teach. He reinforced what my dad said, "You can be anything." He had gone from being an industrial chemist to teaching high school chemistry, and so he had vast experiences that were definitely different that everybody else, because he had applied what he had learned. He was the one that taught me to love research. That was personal but probably carried on to my military career. I would also have to say Dan Rosenthal, the surgeon that I mentioned earlier. He was my major professor for third year and fourth year residency. Because he led by example, other than he was not a physical training kind of guy. [Laughs] He wasn't just a doctor, he knew stuff about everything. We would make rounds on Saturday morning, and then we would play a jeopardy type game and he knew everything, history, other sciences, everything!

As far as military leaders, it is sadly easier to pick out the ones that influenced me on what I do not want to be like when I grow up; those guys are easier to come up with. One who had a positive influence though was [General] George Casey who was my chief of staff at my last assignment in the Army. We established the Psychological Fitness Program for the Army. He came to every class that we put on for a year. He was trying to demonstrate that this was important, and if this was important enough for the highest ranking four-star in the Army, it was important enough for everyone else too. He never had to say it; he just did it! Lance McElhiney who as far as I know, he was a warrant officer, didn't have a formal leadership position, but he led people everywhere. Lance could make people walk through fire, and he was an incredibly talented pilot. His greatest talent however was motivating young pilots.

What motivates you?

I guess what has motivated me a lot in my life is that I hate wasting opportunities. When opportunities come along—it's killing me right now to not be taking advantage of the G.I. Bill. I have more degrees than anyone I know, frankly, so I need another degree like I need another hole in my head, but there is no one to whom I can bequeath it. So, it's just killing me to not take advantage of it. Luckily, I still have 10 years to do something about it. [Laughs] Obviously, one thing that motivates me is the desire to take advantage of opportunities. The other thing that motivates me is challenge and looking for challenges that I can meet. Very often they have nothing to do with competing with someone else for something; it's personal challenges. I like the doing.

Will you give us an example of the most difficult professional decision you had to make and how did you work through it?

It's funny, I don't find decision making very hard.

What training and education did you receive that you felt prepared you to be a commander?

I went to the pre-command course like everyone and what I remember from that course is to call a lawyer anytime I had to make a decision, which I didn't find all that rewarding. So, I think what prepared me the most was just life experiences and being led by people. One thing that I did do, which I thought was very helpful, is when I came back from the Gulf I received a scholarship to the Center for Creative Leadership. When they polled the people that worked for me, I found results very enlightening. What I learned from that experience is what everyone should know. Do not treat people like you want to be treated, treat people like they want to be treated, because what I want and how I prefer to work is clearly not the majority. When I thought I was giving them freedom and trust, they thought I was ignoring them. When I felt like I was being micromanaged or untrusted, they felt cared for. So, I learned to pay more attention to people.

In your opinion, what could be done to help prepare military officers better for leadership success?

I am not convinced that formal education will make better leaders. I've gone to all the schools that they told me to go to, and I don't see that I came out much differently. I see that same thing in my peers. There's lots of things to learn in those schools, but I'm not convinced that leadership is one of them that you learn in a classroom. In fact, I'm pretty confident it's not. I realize that's not the party line. I've pretty much always been honest even when it went against popular opinion. What I did do was come up with an acceptable solution with the Army. I said, "Look, everybody has opinions, and not everyone should be allowed to spew theirs forth, so I will talk about nothing that I don't have any more insight into than the average guy." So, I didn't talk about a lot of things that I didn't have any more knowledge than anyone else. I did talk about things like women in combat, the POW experience, military medicine. I'll talk about things I know about, you just have to trust that my truth will reflect well, and occasionally we'll disagree, but I can maintain my integrity and honesty that way.

What worked best for you to resolve conflict?

Listen to all sides, don't make a snap judgment unless someone is shooting at you, and you make a reasonable decision that you can defend logically. One of the things that I learned to say was, "I understand that you feel that way." It didn't make it any more true or it didn't mean that I felt that way, but it did mean that I at least listened to the whole story.

What do you feel as your greatest weakness and how did you overcome it?

Well, I can talk about my lowest character strength, it was probably empathy or compassion. Sometimes when someone would complain because they had to wake up early to take a kid to school. My thought was, well of course you do. What did you think was going to happen? They would fly there. What really happened to me to change that was one of my dogs had a litter of puppies where five of the eight died. For me personally, it was the worst experience of my life. Worse than cancer, worse than getting shot up and shot down, worse than anything that had ever happened to me. What I learned from that is that you should never judge someone else's problem by how you would respond to it. Many people would have had no issue with what happened to that litter of puppies, but to me it was devastating. Since then, I have become markedly more empathetic and compassionate.

As a military leader, how do you define success?

As a military leader, I define success by accomplishing the mission while taking care of the people doing it.

How do you motivate people?

You need to have a mission that you personally believe is worth doing, then the enthusiasm just flows. You motivate them by caring about what they do, besides just at work. For example, I personally don't care one whit about flag football or golf. If I had never seen one flag football game in my life, I would have been just as happy, except for the fact that my guys want to play and if they were going to play then I was going to every game to support them. I went to every golf tournament as well. You have to not just talk about it, you have to do it. You have to care about people and show it, not just talk about it.

Are there any specific leadership lessons that stick out in your career after DESERT STORM that you want to point out to us?

People say that the military and young people have changed, but I don't think they have. I think the environment has changed but not the people. I think, by and large, the military is still made up of people who come in for a wide variety of reasons and stay because they found purpose in what they are doing. The fact that we have a lot of digital warriors now is interesting, but it doesn't mean they're different. The tools we give them are different but not the people. That's my perspective.

How did your experience in the Gulf and Bosnia shape your leadership style?

I don't think it had that much influence. I think experiences when you're not in training, when you're deployed, confirm both positively and negatively what you previously believed.

What do you do now?

I farm, raise cattle, and pastured poultry. We have a few baby thoroughbreds each year, and I show, field trial, and judge dogs, as well as garden. I also have a business where I do psychological fitness training both for the military and civilian organizations.

What do you wish you had been told or had learned at a younger age or rank?

I can't think of anything. Not that I ever thought I knew everything, but maybe knowledge comes at the same time as need. There's no reason for you to know how to take care of a thousand people when you're only responsible for five. Maybe there's things that you learn as you go that you need, but it wouldn't have seemed relative earlier.

One thing that I wish had been clearly articulated to me earlier, we should be teaching everyone as early as kindergarten, because it's as easy as the ABCs. It's that there is an activating event, whether it's a positive pregnancy test or getting shot or having a divorce or graduating from college. There is an event that happens, and then most people believe that you just have consequences, which is just an A to C relationship, whether it's fear or joy or whatever, but it's really not true. There's really a thought or belief about that event that leads you to behaviors or emotions, which are the consequences. This explains why some people who have an identical event, have different responses. Using my own example, I was involved in a vehicle crash while pulling a horse trailer, my truck rolled over and down an embankment, luckily the trailer came free and remained upright. My first thought when things stopped moving and I could see was gratitude. Gratitude that the horses didn't get hurt and my next thought was also gratitude, I'm so glad I was wearing my seat belt because I have sand in my ear and if I had not been wearing my seatbelt my head would have been out the window and might be flat. My third thought was also gratitude, because I was so glad I hadn't brought my dog, he would have been thrown out and probably killed. So, when your first thoughts are gratitude then the fact that this thing happened is inconvenient and a little expensive but it's not traumatic. On the other hand, a friend of mine also had a vehicle crash, rolled her car at the bottom of a hill, but wasn't hurt. I didn't know all of the things that I do now so I never asked her how she felt, but I watched what she did. She wouldn't drive down that same road, it took her 30 minutes longer to go to work and she wouldn't drive at night, in fact, she wouldn't drive at all for about a month and then she went and bought some enormous vehicle to drive. As I thought about this, I came to the conclusion that she must have thought that, "It's a dangerous road. I'm a bad driver, and I wasn't driving a safe vehicle." So, for her, it really had a lot more negative consequences because of those beliefs. If everyone would learn about these ABCs, beliefs or faith in the absence of evidence. Sometimes they're true, sometimes they're not. So, you have to go actually look for evidence to change your consequences. I used to just say, "I'm indefatigably optimistic," but now I know why. Because I instinctively did that. Since I instinctively did it, it never crossed my mind to teach anyone else to do it. You can teach this at seven, you don't have to wait until someone is 45 years old."

Is there anything as a female officer that you faced that you could pass along how you dealt with it to female officers today?

Well the most important thing, I believe, about being a member of any group trying to be a part of a different group, is to identify with the activity that you're trying to do, not your original sub-population. So, whether you want to be a commander or a jockey or a third grade teacher, you have to identify with that activity. At least while you're doing it, more than whether you're a Catholic or a guy or a girl or an African American. Those things may be important to you in your personal life, but if you don't let those things define you professionally, no one else will either. However, if you make a big deal about any one of those things, everyone else will too.

How do you handle major change?

Preparation.

How do you help people adapt to change?

Make sure as you're implementing change, you have a plan and that you have articulated the plan and have confidence in the plan. If you don't have those things don't implement the plan. Make sure you give everyone at every level affected an elevator speech and make sure you are so painfully enthusiastic that they can't help but be inspired. Because if you don't believe this big change is worth it then don't do it.

As a commander what attributes do you find most valuable when building a team?

The ability to articulate the vision of what that team is supposed to do. You can get the team all excited, but without clear direction it's wasted. I have been in organizations where everyone is excited but then they walk out of the meeting, and they don't know what to do.

How do you handle criticism?

Quietly. In other words, I don't reject criticism, but I absorb it and think about it and then choose to act on it. If I say something right away, I know it will be defensive. Even if the criticism is unwarranted, being defensive won't help me, so stop doing that. I've gotten better at that as I've gotten older. I think the defensiveness used to show a lot more, but you know as you become more confident then you're more, I think, open to the fact. Often, it's not criticism that you did something wrong, but that there's another right way. As people mature, they should come to realize that there is almost always more than one right way and there might be a more efficient right way.

Do you have an example of a failure or mistake that you would like to share to illustrate how you dealt with it?

The biggest failure I ever had was not picking up Bill Andrews. So, that was a pretty substantial failure. Eventually, I realized it wasn't my fault, there wasn't anything that I could have changed to make that happen. You still take responsibility for the failure as a whole because you were part of the team that went out to accomplish a mission and you didn't. I eventually had to ask myself how I could make meaning out of this experience. First, I take comfort in the fact that we all thought we were doing the right thing, we believed it, we still believe it, and if we had to do it again, we would. The reason you go on those kind of dangerous missions is not just the one guy you're going to get, although he's important. The real reason you're going though is because there are thousands of other people doing dangerous things who need to have the confidence that if they have a problem, you will come to get them. So even a failed mission is still valuable if it says something about the whole ethos of the military.

Is there any one person that we haven't talked about that that has inspired you the most?

I would have to say that my spouse. Kory is very inspiring. First of all, he inspired me to do something besides just go back into a lab and never be seen or heard from again. He was the one, as we're getting out of med school said, "Rhonda I want to go do something militarily relevant," which in his mind was not going to do research at Walter Reed [Walter Reed Army Medical Center, Washington, D.C.]. [Laughs] That was my plan. [Laughing] I said, "Alright. Well, I'll go do something militarily relevant with you" and it was really fun and rewarding and I never would have done it without him. He is such an extrovert and such a team player and I had to learn those things. He said to me once, "You do the right thing not because you'll feel guilty if you do the wrong thing, but you do the right thing because it's the right thing to do." I think he's lived his whole life that way.

What are you most proud of in your career and life?

One of the things I'm very proud of is my daughter, who is in a good relationship and is a veterinarian, which is what she always wanted to do with her life. I would say the most gratifying thing I have done professionally was running Landstuhl. Over 27,000 people went through that facility from 33 different countries while I was there from 2003-2005. We were able to save lives every day!

Brigadier General (ret) Cornum is the personification of a professional military officer who puts her focus on those she leads. She will be the first to tell you that she has not faced an obstacle that she could not overcome unless she put it there herself. There is no science to making a decision. A leader is there to make a decision and they simply need to make the best decision with the facts they have at the time. She believes that to be a good leader you must trust your people and learn how they need to be led. Anyone who would judge "this book" by its cover would be sorely mistaken. The petite physical stature of Brig Gen Cornum contains the heart and soul of a true warrior. One does not have to spend much time in her presence to see the strength of character and mental fortitude that drives this leadership icon.

■ Interviewed by Maj Stanley Grodrian, US Air Force, and Maj Shannon O'Neal, US Air Force.

CHECKMATE

John A. Warden III

FIGHTER PILOT | GULF WAR AIR CAMPAIGN PLANNER | AIRPOWER THEORIST

Colonel (ret) John A. Warden III is a combat pilot with a diverse and storied military career. Born in McKinney, Texas, 1943, Col Warden received his Bachelor of Science degree from the United States Air Force Academy in 1965, his Master's degree from Texas Tech University in 1975, and retired from the United States Air Force as a colonel in 1995. He and his wife, Margie, live outside Montgomery, Alabama. They have two children, Elizabeth and John. Col Warden founded and heads Venturist, Incorporated, headquartered in Montgomery, Alabama. Prior to retiring from the USAF, Col Warden served as the Commandant of ACSC. As Commandant, he revolutionized the curriculum and teaching methodology, introduced the concept of top-down learning, developed an integrated oneyear curriculum, and made computers an essential part of student and faculty life. When he left, the school had the most intensive and modern curriculum of any military school in the world. Col Warden also served as Special Assistant to the Vice President of the United States, working on improving American competitiveness, speeding commercialization of technology, and promoting quality concepts inside and outside the administration. He helped keep significant manufacturing facilities in the US and was co-author of the Vice President's Manufacturing Technology Initiative, approved by the Vice President and the Prime Minister of Japan. As Deputy Director for Strategy, Doctrine, and Warfighting, Headquarters USAF, he and his five Air Staff divisions created new operational concepts for airpower employment including the Air Option, introduced the two concurrent regional contingency force planning methodology, conceived and wrote the USAF's initial Global Reach, Global Power doctrine, and created and successfully promoted the Composite Wing concept. Most notably, Col Warden developed the INSTANT THUNDER air campaign used in DESERT STORM for the first Gulf War, and established the Checkmate Washington area reach-back operation that supported the war effort. Col Warden served as the Commander of 36th Tactical Fighter Wing in Bitburg Germany, responsible for72 F-15C [Eagle] fighters and nearly 5,000 military and civilian employees. While at Bitburg AB, he developed and refined mass air superiority fighter tactics, conceived and built the USAF's first self-help individual equipment store, and introduced major new command and control concepts. As a student at the National War College, [National Defense University, Washington, D.C.] he wrote an influential book on air strategy, The Air Campaign, *which is still in use around the world and has been translated into seven languages.* The Air Campaign *was the first book since WWII to address air war at an operational level. Col Warden served as a combat OV-10 [Bronco] Forward Air Control pilot in the Vietnam War where he flew 266 combat missions, and as an F-4 [Phantom II] pilot at Seymour Johnson AFB, North Carolina, where he deployed to South Korea in support of operations following the USS* Pueblo *incident.*

EARLY YEARS

Will you tell us about your family and childhood? What were some of your most fond memories from your childhood and did they influence your decision to join the Air Force?

I was born in the McKinney, Texas when my father was in the Pacific in WWII.

He was riding a troop train in Australia at the time I was born. So, I didn't see him until two years later when he got back and left the Army. He was an engineer officer who had gone to Texas A&M [College Station, Texas] and subsequently served a couple years as a reserve officer. His father, my grandfather, was a lifelong Army officer who started in the horse cavalry. After WWI he moved into the logistics area and in WWII, when I was born, he was running a substantial logistics operation in Northeast India. At the same time, my father's brother, my uncle, who was actually a Gathering of Eagles Eagle 20 some years ago [Henry E. "Pete" Warden, 1996 Eagle], joined the Army Air Forces prior to WWII, was a fighter pilot in the Philippines when the Japanese attacked, and served on active duty until his retirement in the mid '60s. He had quite an interesting set of combat experiences as well as in research and development and then went on to be generally known as the man primarily responsible for the B-52 [Stratofortress]. So, on my father's side, my father, uncle, and my grandfather were all military and my father was always interested in military history and talked about it extensively. He did a lot of reading, and I know that certainly made me more than interested in the general field of military operations.

My father was an executive in a large company, so he moved around quite a bit and, in fact, when he came back from WWII, he actually started off working in Pennsylvania. We lived a couple places in Pennsylvania. We lived in Maryland, and we moved around a fair amount which was pretty typical of the time. So, I grew up in a combination of the North, Pennsylvania, the sort of the middle ground, Maryland, and then the South, Texas, but spent a good number of my summers in Texas with my grandparents. I have probably more of an association with Texas than with any of the other particular places even though we certainly lived in a lot of different places.

As I was moving towards making a college decision, I had always had great respect for my uncle and thought that I would like to do some of the things that he had done. One of which was that he attended MIT [Massachusetts Institute of Technology] and had actually left when the war began to appear imminent. He left just short of finishing his master's degree in aeronautical engineering. I thought well, I'll go to MIT, but maybe I can't get into MIT, so I will also apply to Purdue [University, Indiana] with the idea of doing aeronautical engineering. Then at the same time, I had an interest in West Point [United States Military Academy, New York] or Annapolis [United States Naval Academy, Maryland] but most especially in the Air Force Academy [Colorado]. Not just my uncle but my father also really said that if you really want to go into the military, you really ought to go into the Air Force and not into the Army. An appointment came through for the Air Force Academy, and that's sort of the answer to your first question.

We wanted to know about your time at the Air Force Academy. Do you have an overall positive memory of your time there is it an overall negative memory?

My large observation was that I have been convinced for many years that it was probably by far the best possible education I could have gotten to go into what I ended up going into, the Air Force. Because you got all the things that struck me then as well as now as being really important for a professional officer. So, the academic things were simply great. Some of the other things I was never convinced that the whole fourth class system really had any significant value, that it had a carryover, but it was part of the experience. I guess the easiest way to answer is if I had to do it over, again would I? And the answer is "Yes," and there wouldn't be any hesitation at all.

CAREER MEMORIES

What do you consider to be the most formative assignment of your career?

From a formative standpoint, I guess you probably have to say that it was the

Vietnam War, that's the thing that really got me thinking there's got to be a better way to do things. I didn't by any means have the right answers coming out of Vietnam but was well aware that there was a problem. Actually, that tour in Vietnam was preceded by a short TDY of several months in Korea following the seizure of the USS *Pueblo* in January 1968. Although the deployment wasn't very long and there were no extreme consequences from that operation, one of the things that just utterly amazed me was the lack of planning at higher headquarters.

I was at Seymour Johnson [AFB, North Carolina] with the 4th wing flying F-4s at the time. We were in the middle of one of these periodic reorganizations that all Air Force organizations have done. This particular reorganization was called the Enhanced Squadron Program. At the time in an enhanced squadron, the squadron commander owned virtually everything that he needed in order to run a deployed operation. That was the maintenance people, even the second echelon maintenance (not the third echelon, not the depot stuff obviously), the police, cooks, I mean everything. You literally could go from point A to point B and the squadron was full-up, it was ready to go, when somebody sent some bullets and some food then you were in business. So, that was a very interesting experience for me and really convinced me that was a pretty good way to organize.

The wing then, as presumably wings do now, that are stationed in the United States, spent a lot of time practicing mobility operations to be ready to deploy quickly. We got the notification on a Thursday that we were going to deploy to Korea, to Kunsan to start with, and the first squadron left on Saturday, second squadron on Sunday, and the third squadron on Monday. Not only the three squadrons that were there but the squadrons had all been augmented by fill-ins from Eglin's 33rd Wing. Augmented by those F-4s we actually ended up having about a hundred F-4s that went to Kunsan. It was an extraordinarily successful deployment, I mean it was basically flawless the only thing that interfered at all with it was the third squadron encountered a storm in the Pacific so it had to wait for a couple days before it could arrive in Kunsan. The airplanes were then parked wingtip to wingtip on one of the cross runways at Kunsan. So there were a hundred, of what at the time were the premier airplanes in the United States Air Force parked wingtip to wingtip and loaded! They all had various sets of bombs on them, and we all looked at this and said, "This is utterly crazy." This looks to everybody very much like that sort of set up at Hickam [AB, Hawaii] when the Japanese rolled in. But nevertheless, that's the way it was. What are we supposed to do here? There were no plans, none. So, by and by somebody said, "Well I guess we ought to have some plan."

The first plan that came out was that in the event of a North Korean attack or something like that, everybody was going to take off and we all had some sort of a color-code. You were red or blue or black or green or whatever and depending on what code you had that told you to go to an orbit point some place above the field. You would orbit until you got down to bingo fuel, enough to travel to Japan, at which time you would go to Japan and land there. That was it. This was the plan. Well, this wasn't terribly exciting and then by and by another plan came out and that plan was: "We're all going to get ready to make this grand attack on where the ship was" [Wonsan port on the eastern side North Korea]. So, the thing was laid out that basically, we were going to fly in a straight line from Kunsan up there in a very unwieldy formation. We had a number of guys in the wing that had come back from Vietnam. In fact, a number of them had been with the wing when it had F-105s. I mean these guys they knew what they were doing. They said, "Look this is just nonsense we're all going to get killed if we fly this way." So with a lot of effort they got the plan changed to something a little bit more realistic. But the whole point of that story was that my assumption had been that people at higher echelons of command knew what they were doing. It turned out they didn't have a clue, none whatsoever.

This wasn't just Kunsan operations; all that Kunsan had done for years and years was a rotation from Japan. People come over and their airplanes rotated in for four months and then went back. It was a sleepy kind of operation so Kunsan didn't have any contingency plans. Then at the time, there was an Air Division, and they didn't have any plans. The Air Division reported to Fifth Air Force and there were no plans there. I said, "Wow this is pretty bad. This should never ever have happened." So, that was one of the things that got me to really begin thinking, not in any intense way right then, about, "What's a better way to do it." I also developed an awareness that the mere fact that there's a higher command authority doesn't mean they have a clue what they're doing. So, if they don't have a clue somebody else better figure it out otherwise we've got big problems.

That was a pretty short experience and it didn't have any adverse consequences because we never went to war. We were there for about six months and we went back home. So, there were no consequences other than a very interesting lesson. Then Vietnam, in my view, was the fact that we spent an awful lot of lives and awful lot of money and had accomplished nothing and that made no sense. It was another good example of they don't have a clue what they're talking about.

Can you tell us a little about your first-person perspective on your experience in Vietnam and how you came there in the OV-10?

I went to Vietnam in the OV-10 because as I was in Korea on this USS *Pueblo* deployment in the first half of 1968, it looked like, for whatever reason, that the war in Vietnam might end soon. There didn't look, at the time, like that there was any good way to get to that war via the F-4 because at the time they were sending squadrons and it looked like it would be a year and a half or two years before you would have any reasonable prospect of getting there. So I began to explore some options and found that if I volunteered to be a FAC, that I could go right away, so that's what I did. Then a bit of luck really, I ended up with the OV-10 which was just then coming into the Air Force inventory and being used in Vietnam as opposed to the O-1 [Bird Dog] or the O-2 [Skymaster]. The OV-10 was a lot better, more fun, kind of an airplane to fly. So I went through the training course at Hurlburt [Air Field, Florida], it was a two-month course, if I recall. Then after that off to Vietnam, little bit of an in-country introduction (a week or two course at Cam Rahn Bay), then got assigned to a FAC detachment at Tây Ninh which is up to the northwest of Saigon. There I was stationed with the 2nd Brigade of the 1st Air Cavalry Division. We had our own quarters, but we used a lot of the Brigade facilities and flew the majority of our operations almost exclusively in support of the Brigade. There were seven, eight of us I guess, that were flyers and then maybe four, five, or six maintenance guys; crew chiefs basically. There was no significant maintenance—if you had a problem with an airplane then you would bring some people in from Bìen Hòa or you'd fly the airplane to Bìen Hòa for heavier maintenance.

In the first part of that tour, the 1st Air Cavalry (CAV) was on a fairly aggressive course trying to expand out and to secure areas leading up to the Cambodian border which was actually on two sides, it went around that way. The 1st Air CAV guys were interesting—we really enjoyed them and worked quite well together. We got to about probably June or May perhaps of 1969 when the President had announced a new policy which, at the time, if you were on the spot seemed unfortunate, but strategically it was probably right. The essence of this was we're going to stop pushing and we're going to go into more of a defensive period to reduce casualties. We were simply not going to win this war, and we were going to reduce the loss of people and airplanes. Which, you know, if you're out there on the front well then you hate that, but if you're up at higher headquarters and really looking at it, it probably was not a bad decision. I didn't like it at the time because I thought we could do a heck of a lot better and there were a lot of things that could be done. So, there was an opportunity for me at that point to go over to fly out of the Nakhon Phanom in Thailand. I took that opportunity and am very glad that I did because that war was a completely different war. That war out of Nakhon Phanom and also we had an attachment out of Ubon [Thailand] flying over the Ho Chi Minh Trail. It was almost no close air support; rather it was all real straight interdiction. So, I had two different sets of learning, although nothing on going north so I missed that part of the thing but never-the-less, conducting operations against traffic coming down from North Vietnam was pretty interesting. So that was two different wars, I mean they were dramatically different wars there was no correlation between them.

Are there any particular sorties stand out to you? We read that you came back from a couple of sorties pretty well shot up?

Yeah, probably the one that was most interesting was flying over the Ho Chi Min Trail down a little bit farther south, and I got hit by probably 23 millimeter. It blew off a pretty good chunk of the tail and I thought, "Uh oh, I'm going to eject from this," which was not at all a good idea because I was flying over a place that I knew was just infested with North Vietnamese and the Laotian forces. So, that was not a good idea at all, so I started playing with the airplane a little bit. I found that if I flew too slowly that it was not controllable. If I flew too fast I could see more of the tail coming apart. I finally ended up with sort of an intermediate speed which wasn't terribly fast, certainly by jet airplane standards. I flew back to Ubon, I was flying out of Ubon that day, and landed at probably two or three times what the normal landing speed was for the OV-10 which was not a big deal in the OV-10 and the Ubon runway was probably 10,000 feet. By flying at that speed, the airplane was not likely to quit flying as I was coming down. That was an interesting flight.

Of the missions, the particular sorties that were especially interesting [and] probably the one that stands out in my mind more than anyone else was Christmas Eve. We were flying over the Mu Gia Pass at night. I flew mostly in the day, but not exclusively and we had some of the OV-10s that had been modified. The OV-10 is a two-cockpit airplane and you could fly it from the back where it had a stick and everything else but some of the airplanes had been modified by cutting a hole in the floor of the rear cockpit and then mounting a starlight scope. We had a handful of navigators that were part of the squadron and they would fly in the back and they would observe. So they were back there looking through this starlight scope, and it was by today's standards quite primitive, but it was a heck of a lot better than flying around trying to look out and see something at night. So we were flying around Mu Gia and the fellow in the back seat says, "We've got some trucks down below us." So, as we made a couple more passes around, we discovered that, in fact, that there was a whole convoy, probably 25-30 trucks, that had just crossed the border from North Vietnam into Laos and probably at that time was stretched down by a mile. So fortunately, we found that there were some fighters that were pretty close and we brought them in and succeeded in hitting the front of the convoy and the back of the convoy. So now these guys are boxed in because the road was narrow with jungle on either side. The North Vietnamese were pretty darn competent in keeping roads open but they weren't four-lane roads. So it wasn't really easy to turn around, it wasn't easy to get off the darn road. Over the course of the next hour or so we ended up destroying I think all of those trucks so that was a pretty interesting operation.

The other part of the thing that was interesting in a different way was that interdiction operation saw the first extensive use of electronic detection, called the McNamara fence. All kinds of sensors, sometimes you would fly missions and your entire job was to go down and drop these sensors that would then send up a signal if a truck or people were passing by. It had a big control center at Nakhon Phanom, you could go in there and you could see where things were. It was an early effort but it was one of those things that you learned something about—ok, so it didn't work terribly well then for a lot of reasons but it clearly foreshadowed the fact that there was going to be some significant new opportunities in the future.

What is your biggest lesson from your time as Vice Wing Commander and Wing Commande at Bitburg AB, Germany? What was your overarching philosophy as the Wing Commander there?

When we talk about Air Force organizations, our ability to employ airpower, fly airplanes, get them ready, do the intelligence, all the rest has been extraordinarily good for well over half a century, since before WWII. We do it the best of anybody in the world, so the issue has never been with the ability to fly and to do the mechanical aspects of airpower. What the problem has been is in the sort of higher-level application and not just the strategic effects but even the operational effects. So just to give you an example of what we're talking about, the basic approach to employing F-15s and their predecessor airplanes in Europe, for a very long time had basically been a slight variation on standard air defense. When [the] enemy attacks, scramble a couple airplanes, go shoot them down, come back and re-load.

The problem in my mind was, "Is that the kind of war that we were facing or are potentially facing in a European conflict?" And my presumption especially after having written *The Air Campaign* and read a lot of stuff about air operations was, "No, it's not right because what we knew, with near certainty, was that if the Soviets attacked they were going to be attacking with masses of airplanes." So, if you scramble twosomes, they go out there and get destroyed. They might shoot down two or four or six Russian airplanes but it's kind of irrelevant because there are so many of them that the rest get through. So, one of the things that I was especially interested in doing was to try to get USAFE to think about the employment of air superiority airpower [F-15s in this case] in a different way. My larger idea on this was that if we are going to deal with a large Soviet attack then we need to meet it with a huge force. This is outside of what you could do against the enemy airfields which was outside of the purview of an F-15 air superiority wing commander at the time. You could think about it but that was a different set of issues. So, one of the things that I really wanted to do was to get, first of all the wing, able and comfortable with doing mass air operations—squadron size and then eventually wing size and to do it very rapidly. Then to get that accepted by USAFE command authorities and eventually by Commander in Chief, European Command (CINCEUR). So, that was probably my number one objective for the wing. I mean on a broad scale and obviously, there are other things that fell under the way that people were living and how you can improve the quality of life and all the rest. But the big objective was to figure out a way that the wing could be dramatically more effective in combat operations.

What was the reception of your ideas on combat operations and were any of them enduring past your time there?

I know some of it was accepted but the reception, as is not unusual, varied significantly. The Vice USAFE Commander was a huge supporter, he agreed completely, and really helped. The USAFE commander was not enthusiastic, so I don't think that he liked it but didn't actively object. It obviously survived somewhat because the wing used some of it, three years later, when it went to the first Gulf War.

Were there any major lessons learned or takeaways from your time as wing com-

mander in more general sense?

I tried to do a lot of things. We had 90 some different things that we wanted to do, that, for the most part, were relatively small and a handful were in the air employment area. But a lot of them had to do with quality of life, efficiency, effectiveness, etc. One of the things I think that I learned out of that was that you could have a lot of good ideas and I think the majority of those were actually pretty good ideas. But the majority of them didn't get executed, and some of them created some real consternation. So, the lesson that I took and have really in many ways incorporated into our strategy education is this idea of planning in the open. What that really means is that if you want people to do something, you must figure out a way to get them to be participants in its creation. You can give people a list of things to do, and you can explain it to them, and they could make perfectly good sense, but if people didn't create them then their default mode is to be skeptical at best and more than likely to simply not do it. It wasn't as if we didn't sit down with the staff and talk about all of it and so on but that's a different kind of a thing than leading them through the process which would allow them to really feel a part of the creation and then to take up an ownership role in it. So certainly, in retrospect, that was a significant mistake.

Do you think that could have been applicable to the USAFE commander at the time—to somehow use that planning in the open concept to make him more receptive to your ideas?

Yes, I think so. There were several things that we were trying to do which made perfectly good sense but had the ability to create a lot of opposition out beyond the return that you would get even if that was the case. One of them was the elimination of reserved parking. We've all seen it, you go into someplace and here are a dozen places that there is no car in. Well, why is there no car in it? "Well, because that's reserved for this guy." "Well, where is this guy?" "Well, he didn't come in today," or whatever. So, there was a business book that I had read called *Thriving on Chaos*. One of the ideas they talked about in the book was to eliminate reserved parking, because it's more efficient and surprisingly, there's a higher probability of finding a parking place if you don't have reserved parking, than if there is. One of the examples from the book was that somebody at a mid or low level comes in at six thirty in the morning and has to park out in the south forty someplace and when he gets up to the front of the building there's a whole row of reserved parking spaces with nobody in it. They start to think, "Who are these guys, I'm here working" and so on. So, I said, "Well, ok, let's eliminate the reserve parking." Well, it turns out that there were a lot of people that really connected their prestige with reserved parking. Didn't make any difference to the position, if you said, "Well you're not going to have that anymore" and even say, "I'm not going to have it either." It was one of those small things that people really resented. That was probably manageable and it was actually moving along pretty well. But then probably a significant error was the first visit that [Gen James] Kirk [USAFE/CC] made to Bitburg. I said, "Well we're not going to change anything." Somebody said, "Do we want to have a reserved place in front of the officers' club where we're going for lunch?" I said, "No, I want him to see exactly how we're doing the thing." So, we had to drive around a little bit to find a place to park and he was just incensed with that. It was not something that I had prepared him for, hadn't talked to him about its advantages; hadn't even thought about it. It never occurred to me that he would particularly care one way or another, but he did. So, you can do a heck of a lot more than one thing at a time, but you've got to think through it a little more than I did. Think, "So, what are the likely reactions to some of these things?" Even if it's a really good idea, is the cost of the reactions worth whatever it is that you get out of it. It's not an easy thing to think about in advance.

It has been noted that you reinvigorated, as it's been termed in one book, "a stagnant culture" at Air Command and Staff College when you served as the commandant. What did you see that needed to change and how did you do it?

I was working for the Vice President of the United States when I received a call from the Vice Chief [of Staff of the Air Force], Mike Carns, who asked me to come chat with him for a little bit. So, I went over, and he said, "I would like you to go down to Command and Staff College," and he said, "I don't think it's working very well. I don't know what's wrong with it but would you go down and fix it?" So, I came down without any specific thoughts as to exactly what needed to be done. I had never attended ACSC. I'd done it by correspondence. I never attended any of the schools at Maxwell. I didn't know specifics, but I basically knew what was being taught. And at the time—grossly over-simplified—the curriculum had been refined over 30 or 40 years to match the pace of the Cold War. Although everybody thought the Russians were really dangerous, which they were, they were extraordinarily slow-moving and bureaucratic. If we think we're bureaucratic, those guys were really good at that. So, it was not difficult to keep up with them. We had been talking about the same war for 30 years and hadn't changed a bit.

In fact, while I was at ACSC the Air University commander was invited over to visit the Soviet— well, at this time the Russian—Command and Staff College equivalent. So, he took three or four of us over with him and as part of the tour, we went down into the General's staff headquarters in Moscow. This general staff headquarters was a six or ten-story building up above and then it went down, maybe that far below. At any rate, we got in elevators and went down and eventually came out into this large briefing room. There was a stage up in the front, and there were seating areas and so on. And a good chunk of the thing was occupied by a terrain map of Europe that basically ran from—well, it started roughly on the Russian-Polish border and then went to the English Channel. There was a Russian colonel, a general staff colonel, obviously a very smart, very sharp guy that made a presentation on how the Soviets had intended to fight the war with us in Europe. As I'm listening to this—I mean it's really fascinating—but as I'm listening, I'm thinking, "Holy smokes. They were planning to do exactly the same thing that they'd done in WWII." They had some slightly different airplanes and new tanks and so on, but methodologically, it was exactly the same. So, as I'm listening to that and making that observation, and I'm also thinking, these guys would have been easy to stop. They didn't have a chance if we'd have done it even remotely right. I only mention that because although we didn't know that exact detail, we'd been thinking about the same place, the Fulda Gap, for all this time, and the Soviets hadn't changed very much in any sort of a significant way.

The curriculum at [Air] Command and Staff College kind of matched that. There would be a little reading that was three or four pages cut out of a magazine article, it was passed out, and the seminars would sit around and talk about it, with people exchanging their views with no right answer, no wrong answer, or anything else. My conclusion was that kind of an approach was simply not going to work for the world in which we then found ourselves in which we didn't know who the opponents were going to be. We knew that the technology was moving very quickly, that it was going to be very difficult world, challenging, and that if Air Force officers—obviously others, too—were going to do their job, they needed to be able to think strategically about what it was that they were doing and have a methodology of thinking through problems. We couldn't even tell them what the problems were yet. They had to learn a methodology.

So, in essence, that meant that we needed to have a new approach. My thought on the thing was that the basic methodology that we had used in the first Gulf War was pretty decent from a strategy standpoint. That approach began by asking, "What did you want Iraq to look like when the war was over? What kind of a future do you want to establish for any kind of an operation? What are the primary things that you must put your efforts against, centers of gravity, systems and how much can you compress the time to do it? Then how are you going to get out, whether you are winning or not winning?" So, those things, then, seemed to me that they worked reasonably well with the first Gulf War, and that should be the underlying part of the curriculum. So then, with that as a general idea, you started talking about it, first of all, with the immediate staff, and then very rapidly moving down into all of the staff, which I guess there was about a hundred, or something like that. So, you start bringing them all into the auditorium and say, "Look, here are some thoughts. What are you all thinking?" You have lots of significant debate on some of it, some of it kind of violent, but after a few months, we have sort of a general agreement.

Then there were a couple three people that were really sold on the idea—of one of them was the guy that became the dean, an officer by the name of Larry Weaver. I said, "What we need to do is we need to have a pilot course of some sort." So, what we decided to do was to create what, at the time, we called the Air Campaign Course. That was going to be a two-month program that would take place in the middle of the year starting right after the second semester started and it was going to be entirely voluntary. So, we roughed it out, started putting it together, then asked for volunteers from the students and ended up with a little over a hundred volunteers for it. [We] ran that for two months and by the time we're halfway through it, we said, "This is working. This is what we're going to do next year," and then began creating the whole new curriculum, which, among other things, included everybody getting computers and a hundred books and so on. Mechanically, if I were doing this again, I would approach it a little bit differently, but basically, it really was planning in the open. It was bringing people in, and giving everybody an opportunity to talk about it, and not shutting down anybody's arguments and all the rest. We even invited students to participate, and a few did, but not very many.

GULF WAR ERA

So this brings us to the Gulf War, which has a number of questions we wanted to ask you, but is there anything else in your career besides your time at Air Staff as Deputy Director and your experiences planning for Gulf War, that we didn't talk about?

Maybe a little bit more on the time leading up to the Gulf War, because it was very important to the Gulf War itself. What we were really trying to do on the Air Staff was to get airpower seen and employed in a different way. To see it employed at a national level more effectively. And that, in turn, meant that there had to be some doctrinal changes in the Air Force as well as higher echelons. For example, the Office of the Secretary of Defense put out planning guidance, which was one level below the White House, which was called Department of Defense Planning Guidance. To simplify a lot, the Planning Guidance basically considered two options for a war against the Soviets. The one option was that you send twelve divisions to Europe really, really fast, and then, by the way, some airplanes that go along with it and then fight a colossal mano-e-mano battle over the Fulda Gap. Well, suppose you don't get there in time. So what's the second option? The second option is the sea option, which was that the Navy would pull back into the Greenland, UK, and Iceland gap and would basically defend against the movement of Soviet naval forces, submarines, etc, sailing towards America. At some point, something happened that was never quite clear, but maybe tilted the Russians to sue for peace, or whatever. And so one of the things that we were trying to do, was to answer the question, "Where's the air option?" Well, there was no air option. Clearly, the air option was, "Don't worry about sending divisions to Europe, they're not going to do you much good anyway. Send every darn airplane that you can, because we can employ those against the Soviets in a whole lot of ways that are going to be dramatically more effective." So that was a significant chunk of what we were trying to do with the six or so Air Staff Divisions that I had.

Goldwater Nichols had been in effect since '86, so that had been for three years, but it really hadn't changed a lot. So the service staffs were still extraordinarily strong. And the Joint Staff was—I would use the word weak simply in juxtaposition—but it wasn't weak exactly, it was just that the service staff were very strong. And the Air Staff, particularly in plans, in XOX [Directorate for Warfighting Concepts] and its subsidiary areas, was encouraged to go all over Washington [D.C.] and talk to everyone. We brought people in from the newspapers and went to think tanks, made talks, went to the White House, and all kinds of things like that. So, we were all over the place. And the idea was to try to create the idea of an air option and a better understanding of airpower and then sell it so that it actually got incorporated into the next Planning Guidance. So, that's interesting in itself, but in talking about the air option you obviously have to start moving down to detail. So, as you're moving down in detail, a lot of those details then became as applicable to a war someplace else; for example, in the Gulf, as they did to Europe. And likewise, as we were trying to sell what you can really do with airpower, we had Checkmate run an exercise on applying the five rings to a European scenario. And we talked about that a lot, with probably a total of about a hundred officers in the Deputy Directorate. And in staff meetings, rather than saying, "Well next Monday there's going to be a whatever," we talked about the substantive issues. So, we put together the draft of the five rings and then circulated it. And every week in the staff meeting we discussed it and refined the concept. So after a while there are a lot of people familiar with the Five Rings. Not everybody agrees, but it's out there, so we decided to do a test against the European scenario.

I suggested to the Checkmate Division that it choose something in the second ring of the Soviet Force in Europe because we wanted to keep it, at that time, focused at the operational level as opposed to up to a Moscow level. The starting idea was to see if fuel would be a good second ring center of gravity. So the Checkmate people went off to talk to the intelligence people and they came back fairly shortly thereafter and said, "The intelligence guys laughed because fuel can't be a center of gravity." And we said, "Well, why not?" "Well, because the Soviets have these enormous underground storage facilities all underneath the Black Forest, and it would take six or eight months to blow them all up, and by that time they'd be on the channel anyway. So, that's silly." I said, "That doesn't sound right to me. Let's go a little bit deeper into it. How does this fuel thing really work?" Well, this is a pretty useful lesson in and of itself, because when you look at something at sort of this macro level, it can seem to be one thing. Then as soon as you look at the next level down it can be something completely different. So it turned out—yeah, in fact, the Soviets did have six-months' worth of fuel in underground facilities. But the way that they had them organized was in depots, in twenty-five separate depots, that stretched from the North Sea down to the Swiss border. Ok, that's interesting in itself, so there are twenty-five separate things; it's not just one huge mass of stuff that's under some trees someplace. But then the next thing we found out was that the way that they had set them up was that there were fuel lines that came in from the rear areas. Trains or pipelines, they filled up the depots, then tactical pipelines and over-land fuel trucks went off towards the front. The depots were organized in such a way that they were assigned responsibility for the corps—or the "front" using the Soviet terminology of the time—the corps was what they supported. There was no provision for moving this stuff any other way such as laterally to other corps. It only could go one way.

The Checkmate guys knew Soviet doctrine was to advance at a certain rate, say twenty miles a day or something equally impressive, and they needed a certain amount of fuel in order to do that. Ok, suppose that you could decrease that flow. Either slow it down or stop it. What would be the impact on a Corps commander that's being supplied by it? Well, any Corps commander in his right mind, as soon as he hears that his fuel has stopped, he's going to slow down. He's got to. And there are some people that say, "Oh, he'll just go to the gas station and use the credit card." C'mon, get real! You just can't do that with that volume of fuel required. So, it meant that that even if you didn't shut the fuel off completely, that a reduction in flow would slow down the receiving corps substantially. So now here's this corps that has slowed down, and what does the guy on the other side do? He says, "Oh, my flank is now being exposed, so I guess I need to slow down too, because I don't want to get way ahead of this other guy." So it became very clear that an attack on those fuel depots, on a depot, would have a significant impact. Obviously it's not going to stop the entire attack, but you could measure pretty clearly what it was going to be. The next question then was, "How would we go about attacking it?" And without getting into too much more detail, it turned out that not using precision weapons because we didn't know a lot about the status of precision weapons in Europe at the time we were doing this thing—if you send a hundred F-16s, they should be able to shut down a depot. That seems like a lot of airplanes, but it provided high probability using standard analysis from the Joint Munitions Effectiveness Manual (JMEM), so analysis inflated reality. All put together this looks pretty impressive. The next step was to set up for a handful of the Checkmate guys to go over and brief USAREUR, the Army commander in Europe. His name was General Crosbie Saint. His reaction to the briefing was, "Wow. If you can do that with airpower, that's what I need you guys to do. I can handle the tanks when they get to the front, but if you can reduce their speed and you can reduce their number, this is excellent. This is exactly the way I'd like to see airpower used." And then he said, "Now would you all mind doing the same exercise for the dispersal of the Soviet reserve ammunition?" Of which there was an extraordinary amount buried all through East Germany. And we said, "Sure." But then the Gulf War came up, the Soviets fell apart, and nobody cared about them or the Fulda Gap afterward.

So, long story, but I'm simply saying that we were doing this exercise with the Five Rings, not against the Iraqis, not against the Soviets in Iran, but doing it against a scenario, which was well accepted in all of its details. Numbers were theoretically and I think actually pretty well-known as to what everybody had in Europe. So, by going through that exercise, then we get into the real world, then we've actually had some very decent experience in using the methodology to do a targeting problem. So, that larger thing, that top-down thing you're working on, the air option, leads to better use of airpower, to better success for the country. Then as you start moving further down, you learn about the details, and begin to have extraordinary appreciation on application to a whole variety of different center of gravity target sets. And of great importance, you get a hundred guys that are getting pretty aware of the methodology, know how to use it, and become accustomed to working together.

When the call came in from [GEN Norman] Schwarzkopf (CENTCOM Commander) that we need some help planning this, had your staff been working the plan specifically? Or had you been working some broad stuff and were just generally able to tackle the planning?

Well, so what happened was this: the Iraqis invaded Kuwait on the 3rd of August [1990], and I was on a cruise with my family in the Caribbean. I got back to Washington on Sunday the 5th and called around. [I] talked to a bunch of people who said there did not seem to be any expansive planning going on. The Joint Staff was looking at some "show the flag" kinds of things. Send a flight of A-10s or something like that.

Send a ship to someplace. So, in other words, nothing serious was being discussed. So, on Monday morning, I put the word out in my group for anyone interested to gather in Checkmate where we were going to create a plan to defeat Iraq and win the war. So, we start putting this thing together, using exactly the same methodology. Then the next morning, we had a staff meeting with the Assistant XO [Deputy Chief of Staff for Plans and Operations] as the real XO was out of town. And I mentioned that we'd started this planning. He said, "Great." So that got conveyed to the Chief of Staff [Gen Michael] Dugan, who said, "Super. This is good." And then the next morning, on Wednesday, Schwarzkopf called and said, "Hey, I need some help." He talked to the Vice Chief, actually Mike Loh, since General Dugan had gone off on a trip. And Loh said, "Well, we've already got some guys started working on this, so we'll send them down to tell you their concept." So I mean, that was another piece of luck, because when we started the planning on Monday morning, I said to everybody, "Look, I don't know how we're going to sell this, but let's put the plan together then we'll figure out how we can get somebody to buy it and execute it." Certainly, in retrospect, that would have been, not impossible, but it would've been extraordinarily difficult without Schwarzkopf. And at the time, I wasn't thinking particularly about any way to get directly to Schwarzkopf and to get him lined up with this, so when he called, this was just great fortune. And we went down and talked to him two days later. Within two days, we had the thing in sufficient form—could that we could make the presentation to him. And he was just delighted with the whole thing. So from then—and that actually did break the elastic limits and for that particular thing—there were some variations and so on, that there was never any serious going back on the basic concept although a lot of people certainly tried to reverse it.

One of our questions, and I've not been able to find the answer to, is why do you think GEN Schwarzkopf, instead of going to his own staff in CENTCOM, why would he have gone back to the Pentagon to ask for that air campaign plan, or for some ideas? My understanding is that they were busy doing planning for a defensive type operation, and had too much going on?

I believe that there were several reasons for it. One of which was that GENs Schwarzkopf and Horner had met with the President, with Cheney, the Secretary of State and others at Camp David on the Saturday right after the Iraqi attack. According to a number of sources, their presentations had not received universal appreciation. Schwarzkopf certainly liked Horner, they got along fine, but he had apparently not found that Horner's plan met his requirements. Second, he had sent Horner over to be his advanced Commander on the spot, so Horner was gone, and the CENTAF staff was in the process of moving over, so they had some restrictions. Schwarzkopf was a very smart and very strategic guy, so he recognized that he had a really serious problem. Just to illustrate his recognition of it, when we'd finished the presentation to him later in the week after we had finished and were about to leave, he said, "This is great! This is exactly what we're going to do!" And then he pointed over to a phone on the desk—it wasn't in his office, wasn't his phone—-but he said, "I have been dreading, that at any minute, that that phone was going to ring and it was going to be the President saying, "Norm, the Iraqis just shot some hostages or something similar, do something about it." He said he didn't have anything to do. Nothing. He really had no options. They had sent a ground force contingent, dispatched naval forces, and deployed F-15 air superiority aircraft, but he didn't have any offensive options and the President had said, "This aggression will not stand!" So he's a smart guy and when he didn't have any options, he said, "Where might I find one?" And when he called Loh, what he basically said was, "Look, do you guys have any good thoughts as to how I might use airpower? A strategic air campaign?" Or something like that. And of course

Loh, says, "Yes, as a matter of fact." So it was an unusual thing for a guy to do. It was unusual for an Army guy to do. It would've been unusual for an Air Force guy to have picked up the phone and called the Army Chief of Staff because people normally stay within their particular lanes. But he was quite comfortable going outside that, and I think it was just one of those things. It could have been any of a hundred other Army guys in that position, and that never would have occurred to them.

You gave the great talk to us last November [2016] about how all the briefings went, from the Chief of Staff, to Chairman of the Joint Chiefs of Staff, to SECDEF, and then over to GEN Schwarzkopf, and then finally to Gen Horner. My question is, do you think part of the reception to the briefing, that had been so well-received by everybody else leading up to CENTAF, do you think part of that was because CENTAF didn't develop the briefing?

In part. But the reaction was not out of the realm of what could be expected. People always resent somebody else coming in telling them how to do something, especially when they think that they are experts in it, whether they are or not. I think that Horner was probably a little more than upset that Schwarzkopf had gone someplace else. We had all fought the same war in Vietnam at one level, but what we walked away with were some dramatically different ideas. So in Horner's mind, a significant reason for our problems in Vietnam was because Washington was choosing the targets. But it really was at a much higher conceptual level where the problem really was. So in his mind, Washington is bad and we were the guys from Washington, the "armchair" kinds of guys, as he put it. And I think it was, basically, just kind of a normal human reaction.

So talking about the plan itself. What interests me is the name INSTANT THUNDER. Is that kind of representative of what the plan was? Was it always the name that came tagged with the briefing, and how it was originally conceptualized, or the essence of what was different? You mentioned that the plan was kind of pushing things to a different level, if you will. Is there some aspect that INSTANT THUNDER represents that?

Actually, there were a couple things about the name; one was that it was designed to win the war in a very short period of time—six to nine days. And it was going to happen very, very fast, instant, and then the other thing was that we really wanted to contrast it with the predominant air operation from Vietnam, which was ROLLING THUNDER, which had gone on forever and accomplished nothing. When you get right down to it, other than the last couple of operations which did, but that was a little too late. So right from the very beginning, I think that we said, "Ok, we're going to call this INSTANT THUNDER," which almost everyone in the Air Staff liked. Then we had a guy that had previously been the head of Air Force History, Richard Cohen. I don't remember quite why he was there, but he'd come in to talk about the name, and he said, "Oh no, you can't call it INSTANT THUNDER," and I said, "Why not?" And he said, "Well, because it's not going to be instant and you're setting up false expectations," and so on and so forth. "It's INSTANT THUNDER," I said. I mean, I understood what he was saying, but simply disagreed with him, then INSTANT THUNDER morphed, completely outside my control, it morphed into DESERT STORM, which was close enough.

Going back to open planning, we talked about that several times and how you used that and go back to that and even today. Last November, the talk that you gave to us talked about how open planning really led to success, or maybe, how well the plan worked, that concept of open planning. So my question is, was it just the XOX that was in the open planning, or was it was just Checkmate who was involved in

the open planning, or did it go outside of XOX? Who all was involved?

Well, so for the first two days when we started the thing on our own, before the Schwarzkopf call, it was basically XOX and then primarily it was XOXW [Concepts Division]. There were some other people in there but that was the largest part of it. At that time though, it began to expand. Some of our divisions had some really great relationships with CIA and some other places, so they started bringing in CIA and Air Force Intelligence right from the very beginning, so it was broader than just XOXW. Then when Schwarzkopf called and Loh said, "Okay, this is the number one thing in the Air Force, get whatever help you need." I called SAC and [Gen] Jack Chain was the SAC commander at the time, told him what we were doing and that we needed some help. Very quickly, there were probably a dozen or more SAC guys, largely refueling with some bomber guys. Soon, other parts of the Air Staff and outside agencies began to come in. This expanded the intelligence outreach to other parts of CIA and Defense Intelligence Agency (DIA), and so on, so by the time we had done the presentation to Schwarzkopf, it was fairly broad, although it was still very much of an Air Force thing.

Then after the presentation to GEN Powell, he said, "Well, okay, so this is really good, you guys have done some great work on this thing, but now it's time to make it joint." So we did that, and called the Navy, Marines, and the Army. The Marines and the Navy—again, from the Pentagon staff—they flowed in with great participation and they worked really hard for the next week on the plan. The Army, by and by, sent one guy. He came in, we told him what we were doing, said we needed his help, and he said, "I'll go back and see if we're interested." Never heard from him again. So, they simply had no interest in it. Turned out later, we found out that the helicopter guys were very interested, but they had not been empowered, I guess, to participate. So, we never had any help from the Army, or from Special Operations. We did the same thing a couple times with Special Operations, said, "Here's a target, we need some help on this." And they look at things, I'm exaggerating a little bit here, but they said, in this particular case, which was biological storage, they said, "This is too dangerous. We can't help on this one." Fine, we'll figure it out. But the Navy, the Marines, were great. And actually, the Navy sent a couple guys that—particularly during the war, I forgot when they came back—they were there early on in the planning for the cruise missile things, but then during the war itself there were two senior cruise missile planning guys that operated out of Checkmate and they did an awful lot of the planning for the cruise missiles right out of there. We had a better picture of what was going on than anybody else, any place. And then, very rapidly, after the session with GEN Powell, we started inviting senior people from all around the Pentagon to come in and get the briefing, but sometimes we'd go out to them—the Marines, the Navy, the Army—a little bit slower, but by and by finally briefed the Army Vice Chief and the Chief. Vice Secretary of Defense, Under Secretary of Defense, Under Secretary of Defense for Policy, high up in the CIA, so we really expanded the open planning concept rapidly outside of not just the Air Force, but also outside of the military.

In retrospect, do you think you might have expanded that open planning even further to try and bring in CENTAF staff, even though those guys were in transition and had that stuff going?

We actually did. TAC, which was sort of a surrogate for the CENTAF staff, sent up four or five guys, a couple colonels, and a couple others, and they kind of watched for a little bit and didn't participate very much as I can recall, but watched and then they went back. They didn't have a very receptive boss, so one of the interesting things that happened was that my boss, Maj Gen [R. Minter] Alexander, two star, head of Plans, said, "We need to send a copy of this briefing down to Gen [Robert] Russ

[Commander, TAC]." And I said, "Well, I don't know about that." And he said, "No, we need to do it." So we sent it down, and the next day we got a response from one of his two stars, that was pretty close to verbatim, said this, "General Russ has looked at your briefing and he agrees with everything in it after the last slide." And then he said, "This is how he thinks it ought to be done." And I think I talked about this in the presentation, a real Vietnam kind of proposal.

The senior TAC guys, with some exceptions, were bitter enemies from the very beginning and I think that the conversations between Russ and Horner may have influenced Horner because Horner was a protégé of Russ. I don't know anything more we could have done with TAC or with CENTAF. And in fact, I think as I mentioned that when we got over to Riyadh, we did a presentation to Horner's senior staff—a couple 2-stars, a couple 1-stars, etc. and there, it was like midnight or something—they're delighted with the whole thing, "Wow, this is great. We didn't have any chance to do anything like this. This is really good. I hope you're all going to stay and help us with this." So, the reception at the below Horner level had been fine, there was no problem at all. So, it was that next step up that ended up being a bit of a surprise. And other than maybe handling the briefing a little bit different, I don't know anything I could have done to have forestalled the reaction. Clearly, there had to have been, but I don't know what it was.

So after that, you come back to the States. Can you tell us a little about the role you played, as far as reach back and the campaign planning forward from that point?

Well, so there were a number of different things there, you know. When we first got back, it wasn't clear what was going to continue to happen in Riyadh, even though we'd left the three guys there, how they were going to be used, etc. But within a couple or three days, five maybe it was, talking to those guys, that it turned out that they really had managed to basically sell the idea down at the staff level, so they were moving along with it. And in fact, that Horner was much in the same position that Schwarzkopf was, he didn't have the staff, in his mind, as he said in the briefing, that the thing that obsessed him were all those Iraqi tanks in Kuwait, and that was his focus. "How do I defend?" That was his focus and that had been his focus for a long time. "How do I support the ground commander's scheme of maneuver (there wasn't any ground commander, or ground forces, but nevertheless)?" So, now there's this cadre of people that have this offensive plan, and [Lt Col] Dave Deptula was a pretty good salesman, while Lt Cols Ben Harvey and Ronnie Stanfil were great staff officers who kept things moving along. Dave Deptula was really good on an interpersonal basis, he's really very good, he's forceful, and he works well with people while being brilliant at the concept level. He'd been one of Horner's Instructor Pilots in something, so he knew him from before, so they had a pretty good relationship. So pretty quickly, Horner is not saying, "Well, we're going to do that plan that those guys brought over," but "We're going to do this plan," which is exactly the same plan. With that development, our three officers said, "We don't have any maps. We don't have any intelligence stuff. We don't have anything." So, we start developing this whole contact system, and we've got great relations with CIA, so we got stuff that the DIA and Air Force Intelligence didn't have directly on hand. So, we just started sending stuff over, and obviously you keep thinking about it, "Ok, so we're up to this point, what more could we be doing?" And so we started developing more targets, more concepts. On the one hand, that's the direct connection, but then the other part of it was that there was enormous opposition in Washington to the whole concept. So we spent a lot of time briefing people, talking to people and so on and just garnering support, fighting off attacks, and all the rest of the stuff. It was full-time business, up until the war started, then became 24-hour day business.

So it wasn't back to business as usual, this was your sole purpose in life?

Pretty much. I mean, for a variety of reasons [it] couldn't be completely, but out of a 10-or 12-hour day, I was probably spending on the average, and it ramped very quickly, probably spending eight hours a day on the war planning, and four hours a day on the peacetime stuff. Which by the time the war actually started, the peacetime stuff had probably gone down to 30 minutes a day or if I need to sign something, bring it to me and I'll sign it.

So into execution, you all were still helping to provide targeting, intelligence, all that same type of information throughout the air campaign?

Throughout.

Out of that whole Gulf War experience, both the briefings, your experiences there, and the reach back that you provided, how do you think that whole experience shaped or changed you as a person? In theory, did airpower theory shape or change you at all?

Well, so I think on the airpower theory side, so you learned a lot more things, so you go smarter in it. But, I think at the basic level, it seemed to me that it validated everything that we'd been talking about. So, it worked and it worked well. And because when you're actually doing something, you're going from theory down to details, so now you understand a heck of a lot better how to do it, how to reconstruct it, and so on and so forth, and that experience certainly was invaluable in setting up this operation, this company, [Venturist, Incorporated] which is primarily in the business of helping people learn and develop and execute strategic plans. So the detail that you learned in actually doing something like that was invaluable.

How about on a personal level?

Well obviously, more than a little bit of satisfaction. I mean it is nice when you fight a war, and win a war, and only 150 people get killed on your side. The other side—we cared a little bit about, but not as much.

Looking retrospectively at the briefing, it didn't go the way anyone would have wanted it to. Were there any lessons learned that you took from that, in the way that you would have approached it?

Well, sure. If you knew beforehand what you knew afterward, there would have been some things that you would have done differently. And some of them are pretty simple, like I would have changed the title slide of the briefing. It wouldn't have said, strategic air campaign. It would've just said, Air Operations against Iraq or something like that. And maybe, during the course of the briefings when General Horner pointed out the window and said, "You don't understand that those tanks might arrive in hours." Rather than saying, "You're over worrying about those tanks." I might have been a little bit more, perhaps a little kinder, and little subtler. "Well, you know, those are a lot of tanks out there, but maybe there's a way that we can deal with those." But I mean, it was one of those things. It was a comment that struck me as being so totally out of touch with reality that it was difficult not to react. But that's not a very good excuse.

How different or similar do you think the actual plan in the Gulf War air campaign was to the original plan that you had presented?

At the first level, it was almost identical. At the second level, there were variations. So, for example, well even some of the detail of things ended up being reasonably close, like I told you in the ACSC presentation, Schwarzkopf said it'd be six to nine days to do the strategic campaign. Well, that was based on six to nine days of good weather, which, because at the time we thought we were going to war in September, and the weather was good then. If you correct for the weather days that we actually had in the real war, it actually came out to pretty darn close to nine actual good flying days, so even that forecast was pretty good. The original plan, as presented to GEN Schwarzkopf, did not attack the Iraqi Army in Kuwait, so that was clearly a significant difference, but when Powell brought that up when we did the presentation to him, he said, "Well, I want to attack the Iraqi Army," and I said, "Fine, we can do that." So, we then created what we called at the time the Phase 3 operations, which was actually the air destruction of the Iraqi Army in Kuwait, and sent it over to Deptula for in-theater consideration. They adopted the majority while adding brilliant innovations such as kill boxes and tank plinking. On balance then, the plan again ended up happening on pretty much the time scale that we had forecasted when it was adjusted for the weather variations. The losses were about the same or were a little bit less for the campaign and then dramatically less for the overall than had been predicted to GEN Schwarzkopf and later to Secretary Cheney. I had anticipated that we would lose maybe a hundred airplanes total in the war, and we actually ended up with, what, twenty-five or thirty US, something like about 75 for the total coalition. So that part was off a bit. The part that we could have been done much better, by me, certainly, was to have gotten a better agreement on what the real exit plan was going to be. Which really depended on an agreement as to what Iraq would look like when the war was over, and then how that exit from military success was going to be conducted. That was regrettable. We didn't do well with that. It's hard to get people energized with it, and I finally just kind of gave up.

THOUGHTS ON AIRPOWER THEORY

I'm interested in the same line of thinking about the current definition of airpower and as you talked about: What is it? What does it do? What can it do? We did a quick look up the definition in doctrine and it is defined as "the ability to project military power or influence through the control and exploitation of air, space and cyberspace to achieve strategic, operational or tactical objectives." So, do you think that definition encompasses airpower in the way that you had thought about it and do you think that rings true or do you think that it's missing something?

As far as that goes, I mean obviously, it's right, but I think that what it does is really only address the mechanics. Here's stuff that goes through the air and has an impact on the enemy. Ok fine, we understand that, so what kind of an impact can it really have? Rather than describing tactical execution, my inclination would be more to address what can we produce with airpower and start talking about that. You can use airpower to convert an opponent from one state into a more desirable state. You can convert them a little bit, or you convert them to zero. You have a full range of strategic conversion operations, so I need to focus on the nature of the conversion or in other words, what I want the enemy to look like when the war is over. I will almost certainly employ kinetic and cyber weapons, but those are the tools, not the ends of airpower. It's imperative to start by asking yourself, "What is it that you're trying to accomplish? What strategic conversion are you looking for and why? What is the purpose of the war?" Because once you understand what it is that you want your enemy to look like in consonance with your view of the future that you're trying to create, that then allows you to move down into the five rings, the centers of gravity, the targeting, and actually affecting various elements of the system. But if you don't start with asking what you are trying to accomplish, what does the enemy need to become, then you start off at the wrong level of planning and then go out and do things that probably are executed perfectly well but have no real value. If you will permit me a minor digression, I would say that without question, I think that certainly is the case in the last 10 years in Afghanistan and in Iraq. We have done an extraordinarily good job of dropping bombs on what somebody said to drop bombs on. And here more than a decade later, we are still fighting the same wars with little to show for the lives lost and money spent. I mean that's maybe a little unfair, but as far as I know it's accomplished nothing, because we didn't start with what is it that we want Afghanistan and Iraq to look like at the end of this operation, nor did we even ask if our hearts and minds objectives were reasonable or achievable in realistic time frames. Now, "What are we going to do?" The answer may be to rethink the whole exercise with a clean sheet of paper. Would we get in now, knowing what we now know, if we were not already there? If you don't have a really high-resolution of what you are trying to accomplish, anything you do is a tactical exercise.

Thinking about all of your flying experiences, what most influenced your future five rings model that you developed at the National War College?

That idea developed over time. My general thinking, right after Vietnam, on airpower, was that clearly you could do a lot more with it, but I was thinking more in terms of more airpower and being a little bit more careful about getting air superiority and so on, but not thinking about things at a strategic level. So, after Vietnam I went to Spain and flew F-4s out of Torrejón for four years, which in itself was more than interesting because, among other things, while I was there, I had an opportunity to participate in and then later, be the wing project officer for a couple of deployments that we made to Iran, which at the time was still under the Shah. That was really interesting because the planning was for an exercise that CENTO [which long ago disappeared, Central Treaty Organization] had been conducting but the United States had not participated in it. It was a big air defense exercise called Shahbaz that took place over Turkey and Iran and [was] obviously oriented towards a Soviet attack. So, the planning involved the major participants and the major participants in that exercise were Turkey, Iran, and the UK. Then the United States, this would have been 1972, decided it was going to participate for the first time since the aftermath of WWII. The idea was we would take a detachment of eight aircraft from Torrejón to the Tehran Mehrabad Airport [Iran] and we would participate and be integrated into this air defense exercise.

So, this is my first experience with international planning. Dealing with the Brits, the Turks, and so on, that was kind of eye-opening because you are used to the way things are done in America. There was a major from USAFE who was in charge of the USAFE planning and then he took people from the wing to help him. What we are normally accustomed to is if you give somebody that kind of a job, within a fairly broad range, he can agree on a whole lot of different things and if he makes a decision, well, then that's what it's going to be. Well, one of the things that struck me right away was in our first meeting, which we had in Cyprus at the British sovereign base area of Akrotiri. The Brits, from a staff standpoint, are similar to us, not identical by any means, but similar. They can make decisions etc. We had a fairly senior Turkish delegation, senior full colonel s, but they couldn't make a single decision—not a single one. Anything that came up that hadn't already been precisely laid out they'd say, "Well, we'll have to go back to Ankara to find out what the answer is." I thought, "Hmm this is very interesting. They deal in a different way." We made the deployment to Tehran, participated, and got to know the Iranian Air Force officers really from the chief of staff down. So, you participate there, and part of the flying operation is actually flying up over the mountains north of Tehran and flying over the southern part of the Caspian Sea, which is really kind of extraordinary terrain. It was the first time that I've been across the mountains into that area and at least to some extent began to think a little bit more about, "Ok, so the Soviets are not a whole long way away from there."

In fact, you could almost see the Caspian Sea oilfields at Baku [now Azerbaijan].

Then after I had come back from Spain, [I] went to Texas Tech and did a masters degree in international relations. Spent a year with that and then went to the Air Staff. I was assigned at the time [to] a deputy directorate called XOXX which was the plans part of plans. XOXX then had underneath it regional plans, a Pacific division, a Middle East Division, and a European division. It had an International Division which did diplomatic clearances and things of that sort on a global basis. So I got assigned to the Middle East Division. I had anticipated and wanted to go into the European Division, because I kind of specialized a little bit in European things when I was at Texas Tech, not extensively, but I thought well that would be the natural place to go, but well as is natural, that's not where I went. I went to the Middle East which, over long-term, was ideal from several different standpoints. One was that it was an opportunity to learn a lot about an area in which I simply didn't know very much. At one time, I was the desk officer for Lebanon, was the desk officer for Morocco for a while, desk officer for Israel for a while, and then after I had been there a little bit, started really looking at sort of the wider scheme of plans that at the time were all under the responsibility of Europe Command. [I] began to look at these things and began to be a little bit uneasy with them. Then [I began] recognizing that we didn't seem to have very much of a unified kind of policy towards all of these middle eastern things which were linked reasonably well from Morocco all the way to Iran.

So, I started working on strategy issues and then finally they created a position. I was no longer a desk officer, but was the Middle East strategy officer, so I had responsibility for sort of all of that. As part of that I began to really think seriously about, "So what is this Middle East area? What is its importance?" Then I began to think about the fact that if you are considering the possibility of a war with the Soviets, the assumption was, and it was explicit, that there might be some stirring around between Egypt and Israel or something like that and we might deploy some forces there but the real war was at the Fulda Gap. So everybody would run to the Fulda Gap and then the Soviets would attack right into the heart of what our apparent strength was. I said, "This doesn't make sense. I think the Soviets are smarter than that. Why would they do that?" So what else could they do? Well, as you looked at the way that the forces were deployed and you looked at the geography you'd say, "You know, if I were the Soviets, rather than taking that chance of something that one might not succeed in and is likely to go to nuclear and probably not a good deal for anybody. What if, as a Soviet, I swept down through Iran, captured the Persian Gulf area and then said, 'Ok let's talk about this now. We got all the oil, happy to make some deals you know, etc., etc., let's have an armistice.'" My conclusion was that given the European sensibilities, if you will, and our general approach to things, that would probably be a pretty successful operation and the Soviets then would end up with enormous leverage. So I really got to thinking a lot about that and then saying, "Okay could they really do that?" Then started going through all the intelligence and finding out what kind of ground units, what kind of air units that they had up there in that area that basically stretched from the Black Sea across the Caucuses all the way to what we think about as Kazakhstan and the Trans-Caucuses area and so on. Turned out that they had pretty substantial forces there. Although they weren't category one divisions, they looked like they were more than capable of doing the kind of operation because there was no opposition.

So, I said, "How would we deal with that?" [I] started talking a lot and convinced the Air Force Chief of Staff, that we really ought to flag this as a serious concern for the country. Rather than just focusing everything on Europe, we also ought to have a very serious focus on this big swinging flank kind of an operation that the Soviets might pull off. So, we put together a presentation on it and made it to quite a number of people. Part of it was what could the Soviets do and then the other part was what we can do. It was clearly impossible to send any reasonable number of ground troops that would have any value whatsoever. So, if you're going to deal with this you had to deal with it from an airpower standpoint. "What could we do?" Well, we could, in fact, deploy seven wings to the Persian Gulf area. The idea would be to use that fairly significant fighter force to destroy a Soviet advance, which you could certainly do. Now at the time though I'm really still thinking about this as a force-on-force thing. Strategically this would be a pretty interesting move by the Soviets but the way that you would primarily deal with it would be to destroy their force movements and their attack columns which you could do with near certainty. That then became the forerunner of what later was actually to become the kind of thinking for the first Gulf War, and then it also led to, "So, ok, if we're going to do this, who is going to command it and what are the plans?"

At the time there was a contingency plan that European Command had put together CINEUR and it was just terrible. I mean it was a silly plan that in the event of there being any Soviet, nothing like what I was laying out, but any kind of a Soviet attack, we were going to deploy some F-5s to Tabriz, which is up there in the northwestern part of Iran. "What are they going to do there?" "Well, I don't know," because all of the EUCOM plans, not only for this but also for the central European thing, were all reaction. It was all force planning things. "Okay, so the war starts, now what happens?" "Well, things aren't going to go well, so we'll go nuclear." "What are we trying to achieve?" "Well, we want to reduce the loss of territory to a minimum, then what?" "Then we're either going to go nuke, or the war is going to end." It was nonsense stuff, utter nonsense stuff. At the time, the plans that are put together by the regional CINCs are sent to the joint staff for review. For the most part, those reviews up until about this time had been, "Well, you know, put a comma here as opposed to there" and "We don't like this word, change that word," but there was never anything substantive. If you suggested anything substantive the immediate answer, particularly of the joint staff officer, was, "Well, this is what the CINC says he is going [to do] so you can't question the CINC." "So, why the devil were we spending our time with this thing?" So, we started really questioning the command organization and fortunately had some really good support from the Air Force, from the XOX and the Chief of Staff and so on. So, as we are laying this thing out and saying, "You know this isn't going to work. This is the way you ought to do it." Clearly, you couldn't write the plan for the CINC out of Washington for a lot of reasons, but then the other thing that becomes fairly clear is that CINCEUR has. 1) virtually zero interest in Iran and 2) zero capability because if you really think that the Russians are going to come storming across the Fulda Gap, then that's where you live. So everything is focused on that particular area and everything else is secondary or tertiary or nothing at all. So then that suggested, "Well, maybe we don't have the right command arrangements" and that, in turn, became the basis of what became sort of a Middle East joint task force kind of thing and then in the next year or two that evolved into Central Command.

Ok so, again back to why this long story. My thinking is that you can really use airpower very effectively, but it was simply, you know, send a bunch of airplanes up, drop a bunch of bombs, blow up a bunch of trucks and tanks and so on and so forth and everything will work fine. It wasn't until I got to the National War College and ended up deciding I was going to write a book on airpower that [I] really needed to start thinking about, "What is airpower? What's the purpose of it?" Pretty rapidly, [I] began to move out of the idea of going out there and just blowing up a bunch of airplanes on the other side or tanks and so on and so forth. The book, *The Air Campaign*, talked about the concept of centers of gravity and centers of gravity was a good start on the [five] rings because it was a way to think about different categories of targets as opposed to just going out and dropping a bunch of stuff on a bunch of stuff.

A couple years later, after I had been in Bitburg and then had come back to the Air Staff, Gen Dugan, who was then the XO, called me in and he said basically, "We have a problem in the Air Force. [I'd] like to get your thoughts on it." He said, "Our problem is that we are all thinking alike." His words. "We're all thinking like heavy equipment operators. You're an F-16 guy, what do you do with an F-16?" And so forth. Which was pretty much true, and he said, "I know this is not right. There's got to be a better way. Give it some thought." He had read *The Air Campaign*, and he said, "Let me know what you think." So, as I went back and started thinking about that, [I] said, "Well, what we really need is to be able to explain what you really do with airpower, beyond fly out, drop a bomb, fly back. What is it that you're trying to accomplish?" So, then that forces you to think about, "So what is war? What are you trying to accomplish?" That in turn led into the development of the five rings. [I] realized then, pretty quickly, that I'd really gotten the centers of gravity about right in *The Air Campaign* but had never thought about organizing them in any systematic way. The five rings then provided the opportunity to think about the enemy as a system and now to use that as a guide for your targeting priority. Then, more importantly, [the five rings allowed you] to accomplish genuine strategic things with airpower as opposed to merely to going around there and dealing with the other guy's military forces. So, that's the evolution of the five rings.

What are your thoughts on you and your theories and thoughts being compared to [Gen] Billy Mitchell in terms of both innovative thought on airpower and also being a bit of a controversial figure? I think Colin Powell described you as being "brash." What are your thoughts on being compared to Billy Mitchell in that manner?

Well, I think it's pretty nice really when you get right down to it. So it's very difficult to think of any advances in thinking any place that didn't start from someplace else. Then thinking kind of moves up like that. It may not be quite that straightforward, it may be more like that, maybe like this, but people build, people look at what people before them have done and then figure out if there is a better way to do it. So, Mitchell, in my view, was an extraordinarily important man from several different aspects. One of which was that he really galvanized the country and Washington to think about airpower. Had he not done that and had he not done those dramatic demonstrations against the [SMS] *Ostfriesland* and so on. I don't know that we would have lost WWII, but it sure would have been a much different kind of operation. Because without airpower, WWII for the United States would have been—I can't even imagine it. You could do it I guess, but I don't know how. So, I mean, he really set into motion the things that provided the success in WWII and drove to the establishment of, certainly the Army Air Corps and relative independence even before actual independence. So, from his conceptual view of what you could do with airpower I think was just extraordinary. Now did he get down into the details of strategically how do you employ it and so on? Not particularly, but that's okay because he did something else that got it up to the point where somebody else could think about that.

LEADERSHIP LESSONS

You asked a little bit about views on leadership. An example, it's one of the things you want to get into it some point, but the comment that I would make in that area right now actually goes back to education of officers. If you look at the Air Force Academy for example, and you say, "What's really, really missing?" There are no courses in marketing anymore, never have been, to the best of my knowledge. I mean marketing from a commercial standpoint. How you sell stuff. When you really get down to it, one of the principal jobs of a leader, if the leader is anything more than just a manager of the status quo, is really selling stuff—selling ideas, selling products, whatever it may happen to be. We don't teach that and not only do we not teach it but we actively discourage it with a lot of our military culture. This is something that I was aware of before, but it really became clear pretty quickly as I started getting into the commercial world and had to sell things. So, you if you go up to General Smith, let's say, and you say, "Hey, General, I think we ought to do ABC." And he looks at you, and he says, "Dustin, I don't think so." And you say, "Yes, Sir," and you never go back and see him again! If you're in the commercial world, you said, "Ok, Yes, Sir." Then you walk off and say, "Ok, how can I make him change his mind?" And you come at him from seven different directions until he finally says, "Please, just go ahead." But we don't do that. And so, we impart, we have an idea that a guy that happens to be a superior rank or position knows more than you do, which for the most part, he probably doesn't. He probably actually knows less about any given subject that you know a lot about because he's dealing with a whole bunch of different things and so you're the expert. And yet, he says, "No," you say, "Okay, I guess the General said no, so, therefore...

I think you nailed one of the questions we were going to hit on. We were talking about the idea of leading up which has started to come into the conversation a little more. It's anyone influencing at a higher level of the organization, and my interpretation of it, and that's exactly what you're talking about.

Exactly. I would mention as part of that—so, we use the five rings and centers of gravity in business extensively in the commercial application. Now, some people say, "How can you use something that you used in war and then use it in business?" And as it turns out at a strategic level, business and war are identical, there's no difference. At a tactical level, it's much different. You shoot people in war and in business they want to give you a dollar, but other than that, it's exactly the same. You can teach people that if you're going to get to someplace that's different than today, make more money, whatever it may happen to be. There are two systems that you need to deal with, that you need to analyze from the center of gravity standpoint and then do something about it. The first one is your internal system, that's the organization in which you exist, so, ACSC. You want to do something in ACSC, then you figure out, "What are the centers of gravity in ACSC?" But then when you look at ACSC, right away, you say, "Okay, but ACSC is not independent. There is an external organization that has significant impact in what it does." And at the simplest level, that's Air University, and it extends well beyond that. So, if you wanted to do something specific in ACSC, you have to go convince the Air University Commander, who would be a center of gravity. You would have to change that external system so that it would facilitate what you wanted to do within ACSC. And people, all the time, make a huge error, even people that are project leads, but they put together a really neat little project, and then when they start trying to do it, they find somebody one level up from them says, "You can't do that." "What do you mean? It's a good idea." "Well, I don't care, you can't do that." Because they failed to think about the external system at the same time they were thinking about their internal system and you've got to change them both at the same time. You can't do it sequentially.

Do you think, in terms of leading up, influencing people above you, planning in the open, and convincing them that your idea is a good idea, that is something we should incorporate? Do you think there's a right way and a wrong way to interact or try to influence superiors above you, in terms of leading up?

Well, clearly the answer by definition is, "Yes." Obviously, there are wrong ways to do it. You can't go up and poke somebody in the eye and say, "I want to do this!" They're probably going to say, "Well, I don't think so." So, it gets down to one or two things. One, you must convince them that it's the right idea, and obviously, a good way to do it. Second, is to get them to think that it's their idea or that they contributed to it in one form or another. But as long as they say, "Yes," and mean it, well that's okay. However, you've got to make it happen. If they're not going to say, "Yes." You've got to figure out another way to get it done, which can become quite chancy. But if you want to accomplish something, you may not have any choice but to do that.

Is there one word you can think of to describe the most important aspect of leadership? One word for what's most important to you?

I don't know that I could put it into one word, probably into a few words, but it seems to me that the thing that is essential for leadership, is that the leader intends to and tries to take his organization to a higher place. Higher could mean anything, but to a different place, one that is appealing to the leader, which then he, in turn, makes appealing, sufficiently, to the people [he's] trying to lead. I don't know how to put that into a one word term.

What makes a good commander, specifically the squadron level?

I think that it's probably important to think a little bit about the different kinds of leadership, and by that what I mean is that there is leadership of a group in which you have some degree of autonomy, and then there is heading an organization in which you are given the essence of what it is that you need to do, and you are managing the operation to make it work as well as possible. I think in the majority of cases, that the job of the squadron commander is significantly managing and that's probably a bad word to use, because we know that over the years within the Air Force, I presume the other services as well, that people like to distinguish between leadership and management. Yet the majority of what somebody does, who is in a command position in today's world, is actually managing the things that are going on. So, the management, I think, takes several things. One is to ensure that the organization actually accomplishes what it's charged with accomplishing and the second one has to do with people, and that is to ensure that people are treated justly and that they are assisted to the maximum extent possible in being able to improve themselves and be ready to move on to another position. For the most part, somebody that is the head of a military organization in today's world doesn't have a lot of flexibility to get much outside of that particular area.

Do you think Billy Mitchell's style—standing on the rooftops and promoting those ideas in a controversial and a brash manner to make sure he was heard—do you think that was the right way to go about promoting his ideas?

Well, I would reverse the question. Had he not done it that way, what would the outcome have been? I think the outcome would have been that there wouldn't have been any change. Without somebody really pushing, change at best is incremental, if it changes at all, because it's very comfortable to stay with the status quo. There was certainly no particular interest within the Army or the Navy to get into that sort of stuff. I mean they were perfectly happy, the admirals had some pretty nice boats, and the army guys had some nice looking horses and some other stuff, all perfectly happy. So without somebody getting out there and just driving the change, I don't think it would have happened. I'm not even sure how commercial aviation would have developed. You have to believe that what Mitchell did and what that inspired within the next echelon down in the Army Air Corps that in turn had an extraordinary influence on the development of commercial aviation. Commercial aviation would have developed over some period of time, but it would have been dramatically slower, and it's not clear whether the United States would have been the dominant developer of commercial aviation the way that it ended up being the dominant developer. So if you think that the advent of serious airpower was good for the country, then you must say, "General Mitchell, thank you!" And I don't know any other way that he could have done it. Because he had to move something out of one state into another state and you don't do that by just saying, "Gee, guys do you think it would be nice if we had some airplanes or something?" It just doesn't happen.

You've been retired for a few years now. If you could go back and give 2d Lt Warden any advice to heed in his career what would it be?

I think you have to think about that from two different directions and you'd have to kind-of define career. So, the first thing would be, you must decide what you really want and what means most to you, and then you must pursue that. But at the same time, you've got to be willing to think about how you are going to deal with issues when you know that you're right and nobody else agrees with you. Is that something that you are willing to pursue, and ok, career is not relevant, or is it something you say, "Alright, I can put this aside." Good people will make different decisions on that and whether you can say that there is a single right answer, "I believe you do what's right," but in some cases, that may not always be the case. And in some cases, in fact, it makes sense to figure out a short-term compromise if you are certain, or relatively certain, that a little bit later you're going to be able to achieve what you want to achieve. But that's a bit of a dangerous path, nevertheless it's one that probably everybody has to take to one extent or another. Some battles are not worth getting into. But it's also dangerously easy to get into the mode of saying, "Well, let's not do any battles."

Col Warden's accomplishments and experiences are inspiring and absolutely thought-provoking. Had he not had the assignments throughout his career leading up to the Gulf War, would his theories on airpower have been different? What would the war have looked like if he hadn't been involved in the planning? To our good fortune, his ability to layer his beliefs on leadership with his ideas on airpower, which he developed throughout his career, led to the development of the most successful US air campaign of modern times. We owe a debt of gratitude to Col Warden for solidifying airpower's place in US military strategy and furthering our understanding of its capabilities and potential.

■ Interviewed by Maj Adrienne Stahl, US Air Force, and Maj Dustin Wade, US Air Force.

John Warden, Margie, Betsy, and John IV at at Torrejón AB, Spain upon return from Incirlik, AB Turkey deployment.

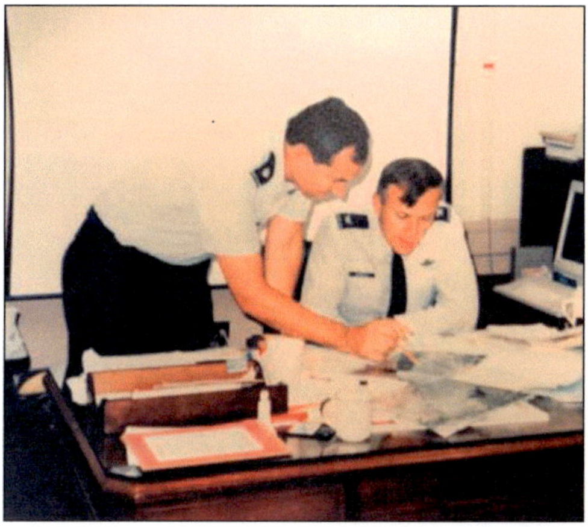

Lt Col Dave Deptula and Col Warden, early Gulf War planning in Checkmate, Aug 1990.

Briefing Secretary of Defense Dick Cheney, late Jan 1991.

(L-R) Ben Harvey, John Warden, Dave Deptula, Ronnie Stanfill, Royal Saudi Air Force Headquarters, Riyadh AB, 19 Aug 1990.

Checkmate planning participants, taken after the Gulf War at the River Entrance of the Pentagon, 1991.

335TH FIGHTER SQDN
CHIEFS

Thomas E. Griffith, Jr.

F-15E WEAPON SYSTEMS OPERATOR | DESERT STORM PRISONER OF WAR

Colonel (ret) Thomas E. Griffith, Jr., F-15E weapons system operator, was shot down during the early days of Operation DESERT STORM. After a successful ejection, then Major Griffith, and his pilot Colonel David Eberly, evaded the Iraqi forces for two days until their capture. On 25 January 1991, CNN broadcast an Iraqi interview with Major Griffith and two others providing proof of life. Major Griffith was held as a POW until his release on 4 March 1991. Influenced by his father Griffith had ambitions of becoming an aviator. The path, as he saw it, was through the United States Air Force Academy. Originally from Pittsburgh, Pennsylvania, Col Griffith graduated from Pope John High School in Sparta, New Jersey, and attended the United States Air Force Academy Prep School from 1974—1975. Graduating in 1979 from United States Air Force Academy with a bachelor of military history, Griffith was able to achieve his goal of a career in aviation by entering navigator training at Mather AFB, California. Griffith was assigned to Homestead AFB, Florida flying the F-4 [Phantom]. Following the assignment at Homestead AFB, he continued to improve proficiency and tactics at each of his units to include Osan AB, Korea; Ramstein AB, Germany; and Seymour Johnson AFB, North Carolina. While at Seymour Johnson AFB, Griffith transitioned to the new F-15E [Strike Eagle] aircraft. Griffith deployed as the weapons system operator for the F-15E as DESERT STORM began. On one of the first combat missions of the war, Major Griffith's F-15E was struck by an SA-2 SAM forcing an ejection at 25,000 feet and 500 knots. Safely on the ground with Col Eberly, Major Griffith was able to avoid detection and capture for days. During this time he was listed as "Whereabouts Unknown." Upon reaching the Iraqi border, Major Griffith and Col Eberly were captured near the Iraq-Syria border. Major Griffith was eventually released when the war ended. After the war, Major Griffith returned to flying and had a highly successful career. He was a DG of ACSC and went on to earn a PhD in History from the University of North Carolina. Utilizing his PhD, Col Griffith served as the Commandant of the School of Advanced Air and Space Studies (SAASS). He also served in staff positions as the Special Assistant to the Chairman of the Joint Chiefs of Staff and the Commander of the 39th Support Group, Incirlik AB, Turkey. Colonel Thomas Griffith retired on 1 September 2008.

Can we start with your basic background information, starting with the Academy?

After I graduated high school, I went to the Air Force Academy Prep School for a year and then went to the Academy and graduated in 1979. Then I started my flight training and went into F-4s. I was an F-4 backseater. My first operational assignment was Osan AB, Korea. I spent a year there then spent 3.5 years in Germany at Ramstein AB. After I left Germany, I did a one-year assignment at the Pentagon. It was called the Air Staff Training Assignment (ASTRA) program; it's probably still not around. I left there and went to Seymour Johnson. I got there in 1986, so the first two years, I was flying the F-4. Then we transitioned to the F-15E. I was in the initial cadre of people that were going to go for training and then come back to Seymour Johnson. I went and got checked out at Luke AFB [Arizona] in the F-15E, and was back at Seymour when the Gulf War started.

Up to that point, was any of the training that you had accomplished focused on the

Gulf War, meaning did you do any spin-up training?

In August of 1990 when Saddam Hussein had invaded Kuwait, it was just a typical August in Seymour Johnson. It was hot and humid. We were flying regular training missions. We had just gotten Initial Operational Capability (IOC) in the airplane. So, there was a lot of equipment that was not on the airplane then that is on the F-15E now. We were figuring out tactics and everything else. And when the war started, no we didn't have any spin-up because the notification was, "Go home, pack your bag, you're gone." So we all went over, most of us went over. We landed in Oman, and that's when we started planning for how we were going to stop an initial incursion. We were getting used to flying in the desert. Getting used to flying with the weapons that we'll have to use if this ever came to pass. Once that threat abated, and we were going be in there for the long haul, then we transitioned to training in preparation for what would become DESERT STORM. At the time, so about October or so, I actually went back to the States. Me and a couple of other guys had to train, since I was an instructor, we had to get the second squadron ready. We came back to Saudi Arabia at the end of December. From the end of December until the Gulf War started we were practicing air refueling at night, we were flying a lot of practice missions at night using the pods and checking those things out.

Fascinating to hear about the no-notice deployment. Currently, our spin-up training begins

90 days prior to deployment. I've had the ability to know for the past few years when and where we'd go. It really was a zero start. At the time, the force was built that way because we were ready to fight in the Fulda Gap or Korea. Those are always going to be, come as you are. There would be very little, maybe some indications of warning, but not at the length of 90 days. So, yeah it was a different kind of mindset.

As far as my airframe the past few years, we've been able to know when and where we're going. What was the mood at the squadron? Did anyone in the unit expect to deploy?

Well, no. He invaded and there were statements out of the SECDEF and SECSTATE. It was sort of like, Gen Charles A. Horner can address this better than I could. I was at the unit level. They were like this is another one of those crises, and we're probably not going to get involved. Maybe we might get involved. Nobody was really sure we were going to go. Until it happened. They called us in and said, "pack your bags."

Was your squadron informed of what force packaging would be overseas when you got there? Did you know if personnel recovery assets would be in theater? Or tanker support?

When we first got there, it was all flowing in sort of simultaneously, so it was always changing. You know the F-15s from Langley AFB , Virginia went first to Dhahran, Saudi Arabia, when we went in we were third and I think F-117s went in after that. It was a daily sort of thing of what's in theater. At the beginning when we were first there we weren't going to have good information on where the enemy is nor where the friendlies are. You know, you're going to hopefully get some information en route because it is a long flight up from Oman, and we'll get some targeting en route. Nobody was sure. It was a pretty confusing time.

Going back to your background leading up to the Gulf War, how long did you have to plan to fly the sorties in theater? Did you have the opportunity to fly dedicated training sorties or was it all combat?

When we came with the second squadron, we got there at the end of December.

We got there on the 26th or the 27th. The war started on the 17th or 16th of January. So we had two weeks. By that time everyone knew what the day one was going to look like. The first, basically, three days were fragged out. Everybody knew. The first night was going to be the most critical one. There were going to be a lot of aircraft in the air and a lot of things were going to happen on pretty close timing. The next couple nights were a lot more fluid. You know, for us we were going to go in at medium altitude. We knew we were going to go in at night. And we knew we weren't going to have targeting pods, so we'd have the navigation pods, and we were dropping bombs that were just glide bombs. They weren't Precision Guided Munitions (PGMs) at the time.

Emotionally what was going through your head? What were you feeling?

Well, I think the first night everyone felt the same way, which is you don't want to—the first priority is not to screw up. Not many had been in combat before. One squadron commander had been in Vietnam. The wing commander, General Hal M. Hornburg, had been in Vietnam. But that was about it. Nobody else had been shot at. Everybody was confident in our training, confident in the airplane, but there was still a lot of uncertainty about what the Iraqis were going to put up. How well this suppression plan was going to work against the air defenses. There was a lot of, I wouldn't say fear, but I'd say people were nervous. They were equally nervous about not screwing up.

One thing everyone has in common here is you went into situations with a lot of unknowns and there was a bit of apprehension about that. We want to pass these lessons on to future leaders. Here is a model; here is what they are thinking. Why did you choose the Air Force?

Well, I wanted to fly. My dad had gone through aviation cadets and flown F-84s [Thunderjet]. He was an airline pilot. He had flown in the Air National Guard for a couple of years, so he was an airline pilot. That looked pretty interesting to me. It looked like a good way to make a living. I was interested in flying and at the time I was graduating from high school, was the end of the Vietnam War. There was a big draw down. I had been interested in going to the Air Force Academy, and my dad said if I really wanted to make sure I could fly, I needed to go there rather than through ROTC. That was a big part of why I ended up doing that.

Any regrets?

Haha, you mean other than going to the Academy?

What's interesting about that one decision is how it set you on this course. Given that one decision affected the rest of your life.

Seriously no. Like everyone else, it's not thrilling to go through it all the time. A lot of things that you don't have the opportunity to do, I'm still in contact with a lot of guys that I met there, and I wouldn't change it.

It's safe to say your dad was your biggest influence?

Yes. Clearly.

From a leadership perspective, if you had to point to one or two experiences that shaped who you are as a leader, could you?

From DESERT STORM? Let me say the one lesson that I take away from my experiences in DESERT STORM was preparation. Being prepared and ready for what happened. As I talked about before, we were in a situation in which we had to be ready to go right now. There was a big saying post-Vietnam from a guy named Lt Randy Cunningham who was one of the Navy Aces. His saying was, "You fight like you train." So, every mission we were flying you had people that took it seriously. They got in the habit of touching the switches the same way, of being ready mentally so when you went into combat you didn't have to think about it. Of course on a bigger scale, you had things like RED FLAG. It was a surprise when we came down the first night and everyone landed from the wing. General Horner can talk better than I about the casualty estimates for the first night, but it was expected that we'd lose a lot of airplanes from our wing; we didn't lose any. We only lost one plane in total and that was an F-18 [Hornet]. I can remember that all the guys I flew with thought we were as prepared as would have wanted to be to go to war. There are some things that you can't replicate in training, but I really felt like we had done a night RED FLAG, and the red air were F-15s from Langley and they had a hard time finding us, tracking us, and everything else. I knew how well they had trained and I knew the equipment that airplane had. I knew the Iraqis didn't train that way and didn't have that kind of equipment. So, you know I felt, most people felt, we were well prepared for what was going to happen. I think that was true the first night and I think that was true as well when I was a POW. I felt like the training I had gotten was, and I've heard this from other POWs too, that the training we got in survival and resistance training was better than you could have hoped for. I think that really, for me as a leader, the one thing you have to do is spend the time to make sure your people are prepared and ready for whatever the mission is.

Was the SERE (Survival, Evasion, Resistance, and Escape) training you received adequate for your situation?

Yeah, it was more than adequate. Now, in my particular circumstances I didn't follow the SERE training [laughs], which is why I got captured. Once we got captured and we were in prison, I could remember it felt very much like the SERE training, and I did mine at the Academy when I was a cadet. I can remember thinking there were a couple of things that were different. One was that you knew in training that it was only going to last a certain amount of time. You didn't know precisely the hour, but you knew it wasn't going to be months. In the real thing, you don't know how long it's going to be. But I had instructors at the Academy that had been POWs for seven years—I knew it was possible—I didn't want that to happen, but I knew people could live through that. The second thing was that in training I was pretty sure they weren't going to kill me [laughs], but if they did, they'd get in a lot of trouble. Whereas in Baghdad, there were a few times when I thought they were. The third thing is, you know, when they want to hit, hurt you, they are really going to hurt you. I also will tell you that when I was going through my SERE training and I was going through my interrogation, I thought I was doing what I thought I was supposed to be doing by not answering the questions. Apparently, I wasn't because there was an officer observing the sessions who called an academic situation. Now this particular guy's name was [Lt Col] Jon Reynolds. He was an F-105 [Thunderchief] pilot and a history instructor at the Academy. I knew him since I was a history major. He had gotten shot down and was a POW for seven years. He came into the room and he said to me, "Cadet, you can't John Wayne it. You can't just sit there and say, 'I don't know,' because they are going to make you hurt and they are going to make you say something. You had better not John Wayne it." I said, "Yes sir."

It was one of those things that you have in the back of your mind somewhere, but it is amazing when you're in this situation what comes back to you. When I got back, I found his address and sent him a note that said, I just want you to know this made a big difference to me.

That is another point about being prepared. It's interesting because we had a SERE conference about a year after we got out of the Gulf War. It was held down at

Fort Belvoir [Virginia]. The Navy, Marines, and Air Force aviators had been through their SERE schools, but some of the Army helicopter pilots hadn't been. Obviously, some of the other people that had gotten captured hadn't been. Everybody that had been through said there wasn't anything that they hadn't been expecting. Now, some of the things are harder to deal with when you're actually doing them. We had the good cop, bad cop interrogation. We had all the fake news about how the war started. I was ready for all the stuff they were telling me. [The training] was very good.

Can you tell us the story? Can you start at the beginning?

It was the fourth night of the war. The day before there had been a number of Scud missile firings against Israel. We were on the night schedule. I had landed before dawn. I was to fly the next night and go against an ammunition storage area near Kuwait. So, sleep during the day. Get up. Come into the squadron. Everything's changed. Everyone in the wing is now going out west to bomb the missile storage areas. At the time it was the geopolitical thing to do to keep Israel out of the war. We were going to go up there. They were mobile sites, and we were going to drop cluster bombs on the whole area. It was also heavily defended. There were a lot of SAM sites up there. The mission is thrown together, not well planned. We were supposed to have—because at the time we didn't have good internal jamming against the SA-2s and SA-3s—the EF-111s [Raven] were supposed to do stand-off jamming and we were to have Wild Weasels with us too. We takeoff and it's not a well-coordinated mission. We're on the tanker. We get our gas. We're getting ready to go to the target. Just as we're about to go off the tanker frequency to the strike frequency. The Weasels are checking in to get their gas, so really we weren't going to have the Wild Weasels with us. We were going to the target. Come to find out later on, and as we're approaching the target, we get AWACS calls that MiGs are airborne and that they are closing in on the package. There are other F-15s that are doing CAP airborne and they're taking care of these guys [MIGs], I assume. But the F-111s retrograde from the target area. We go into the target area without any electronic support which was not a good situation. As we come into the target area, it is lit up with triple A because they know we are coming, and as we turn into the target area, they [Iraqis] start to launch the SAMs.

I was in the number three airplane. The first, the other guys that landed said they counted twenty-seven launches, so I don't know how many actual SAM launches there were. Whether there was half that or twice that, I have no idea. As we are about eight miles from the target, our radar receiver lights up and it says there is a missile launch. I look outside the right side of the airplane and there are two fireballs, because it is night. They appear to be guiding on our airplane.

We do a bit of an evasive maneuver against them, and as we are heading back to the target there is an explosion on the left side of the airplane. Never been hit in an airplane before but it's pretty clear what happens when it happens. Not really quite sure how much damage there was to the airplane because almost immediately the guy flying the airplane [Col Dave Eberly] pulled the ejection handle. Bail out at night. We were about 25,000 feet going about 500 miles an hour. Fortunately, the ejection seat worked really well. Because we were at such high altitude, you fall on the seat for quite a while, and the ejection seat functions and your parachute comes out. We land on the ground. Because of where we got shot down, we were basically—now keep in mind, I can't see him. It's dark, the only thing I can see around me is the bullets coming up from the triple A and fires on the ground from the explosions. I land on the ground not ready for the parachute landing, so I twist my knee. Almost as soon as I'm on the ground, there's another wave of other jets coming from behind us. There's an explosion. Not that far, it doesn't seem that far. My plan is to get the hell away from the target area. We have a small survival kit called the hit and run kit. I grab that and beat feet. Running away from all the lights, I hid out a little bit until all the bombing stops and to catch my breath. A little truck comes around but doesn't see anything so it goes away. We were very close to the Syrian border, and at the time Syria was helping us out in the war. My plan was to go into Syria, turn myself in, and get repatriated. I started doing that. Also, we had procedures to come up on the survival radio and make a broadcast. At the time we didn't have any...times have changed now based on the survival equipment and communications equipment people have. If this happened again, it would be way different. We basically had Vietnam era survival radios. It was a PRC-90. Horrible.

I've read about them [laughs].

So, there you go. I was making my radio calls hoping someone would hear me. I hear Col Eberly. He is on radio too and we get together. He had been injured in the explosion or the ejection. He is bleeding from the neck, and I try to bandage him up as best I could in the dark. He wasn't really able to walk all that much. We laid low then. As the sun began to come up, we saw a hill in the distance, went up and hid on top this hill. We were up there for two days. The first night after we were shot down, they came back. The guys from the squadron came back and bombed the target again. We made radio calls while we were bombing and one of them heard us, but they weren't able to figure out if we were under duress or anything else.

Is that the first time you made contact?

Yes, it was twenty-four hours and it was the first time they knew we were alive. We didn't hear anything else. We were running out of water. We decided we'll move off the hill at dark and try to go to Syria.

We did that. We were walking at night trying to avoid contact with anybody. We had some near runs, but we didn't make contact with anyone. Sometime at the middle of the night, we get a call from an F-15. He said, "Stand-by, I'm going to ask you a few questions." That was to make sure we weren't under duress. We never get another call from him. We never get asked these questions. At the time we were stopped. We were cold. We didn't have any water. We could see a building in the distance. We should have stayed away from it, but thought maybe there was a water pump near it. It looked abandoned.

We thought the building was abandoned, but it wasn't. It was an Iraqi border post. They heard us. They started firing. We started running away. I was...Dave basically, they, they got him. Then I stopped. I don't know why, but I just decided that if they've got him, I'd better stay with him. They were running after me. We were within a kilometer of the border with Jordan. We were at an Iraqi side of no man's land. The Iraqi's captured us. We thought we might have made it, but when you go into the building and see all the pictures of Saddam Hussein, it's clear you didn't make it.

You actually walked in and you saw...

Well, we didn't walk in. They pushed us in with their rifles. We were there and interrogated there. We were taken into the nearest town, where we got interrogated somewhat there. The next night they blindfolded us, handcuffed us, drove us to Baghdad. We spent the rest of the war in four different prisons in Baghdad.

You were at the border house for a couple days?

Probably about 36 hours.

When you spent the time in prison, what was that experience like? How would you explain that overall?

Well, I wouldn't want to repeat it. We were kept in solitary confinement. In one of the prisons we were in, you couldn't communicate between the cells. One guy, an A-10 pilot Cap Dale Storr, he got shot down and taken right away to prison. He didn't even know there were any other people in there. It was thick concrete walls and steel doors. Solitary confinement. We got one meal a day. Terrific weight loss program. There was the typical, a lot of interrogations, a lot of wanting to get information out of you about your airplane, what were you doing, where were you flying from. Then later on some of the interrogations became a little more philosophical, if you will. Why can't we have nuclear weapons, when Israel can have them? Things like that. It wasn't great.

How do you respond to that? Are you in survival mode or are you resorting back to training?

Yeah, survival mode based on the training I got. I knew to keep active and keep your mind occupied. Try to communicate to other prisoners. In the first prison we were in we could do that fairly easily. Dave was in the cell next to me. We could whisper when the guards were not around. We could see other prisoners. We didn't know who they were. The other thing to keep in mind is that we're Americans, but there's also Italians, Kuwaitis, an A-4 pilot who got shot down. There were Brits. You see people, and you hear them talking, but you're not sure who the hell they are. After the first prison we got taken into the second prison which was run by the Iraqi Intelligence Service. In that prison you couldn't communicate with anybody. You could try and say things, but people couldn't hear through the solid steel doors.

Obviously the whole experience was hard. What was the hardest part about it?

Just not knowing what is going to happen to you. You never knew what was going to go on. One day I get taken out of my cell. We got transported in the trunk of a car. We always got transported in the trunk of the car so intelligence couldn't follow us. We got taken down to a bunker and we were in the hallway of this bunker. They take the blindfold off and there's a camera in front of me. The room is full of men and one guy there speaks English. He says, "We're going to ask you some questions, go over the answers, then we're going to turn on the camera. We're going to go over the same questions again and you're going to give the same answers." I expressed, "We aren't supposed to do this." I tried to bargain my way out of it. I said, "I didn't want to do it. This is against the Geneva Convention. You shouldn't be making us do this." He was giving some answers and went on and on. Then he spoke something in Arabic when the guy next to me pulled out a pistol and put it next to my head. "Either you make the video, or he is going to kill you." "Well,"

I thought, "Maybe the questions won't be that bad. Maybe I'll see what the questions are and see if I could wiggle my way out of them," which I was able to do. They asked, "What shot you down?" We agreed we wouldn't say what shot us down. We agreed we'd say we had a malfunction. We didn't want to give them any intel on what worked. Since Dave and I had time to cook up a cover story, we could do that. I said, "It was an electrical fire." The interrogator said, "It would be better to say I didn't know." So that's what I did.

There was another time we got pulled out. Even though you are blindfolded and handcuffed, you could sense there were other people or Americans there. We got pulled into a room and put in chairs. Someone came in. I had suspicions about who it was. They came in with a rubber hose or belt and started beating us with it. There was no apparent reason other than they just wanted to beat the shit out of us. That kind of uncertainty, you get used to it more while you're there. The uncertainty was the hardest part.

What kept you going?

I drew a lot on my training. I knew guys had been through, and women too, the guys in Vietnam had been through a lot worse and a lot longer. I knew the training we had gotten had prepared us for this and you take it one day at a time. When we got shot down, people handle things differently, I thought if we're out of here in a year we'll be doing good. I had no inkling what was going to happen in the war. I mentally geared up for that.

You mentioned you had to keep your mind busy. Do you remember what you thought about during the periods of isolation?

One of the things I tried to do was exercise. I tried to figure out how long it was around my cell. Then it was about a mile around the block in Goldsboro at Bayleaf. I figured out how long it would take to do a mile and imagine I was at home walking around the block. I would think this far I would be about in front of the Ryan's house right now [laughs] we were neighbors.

I remember the guys during Vietnam would try to remember as far back in their memory as they could. Other guys would try to plan ahead. For me, it was best to think about the future and future trips, rather than recall things in the past. I said a lot of prayers. When things would happen out in the hallway you'd try to listen as much as you could just to figure out what was going on.

Did you have a sense of time in prison?

It was daylight or dark. You knew the passage of a day. Every cell I was in I could count the days and I kept track. I knew it was the end of February or whatever day.

You mentioned your training. In addition to your training were there lessons from your childhood that you think might have equipped you for that type of hardship?

I had a football coach in high school that was a real son of a bitch. Vic was a tough guy, and he made you tough too. I think partly that, but I don't think anything in particular from my childhood.

How did your experience change you after the fact?

I've thought about that a lot. In some ways I can't give a satisfactory answer like I became a better person or anything else. One of the things it did do was it made me appreciate things a lot more in life. It made me not be so nervous about what would happen to me since I thought it couldn't get much worse than that.

When I got back into the cockpit flying, I was flying with a guy on one of my first missions back. We were in the airplane, had gone out, and were coming back. He says, 'So how's it feel, are you nervous?' I said, "No, nobody's shooting at me." It's not like the aircraft is just going to break apart. [laughs]

Going back to the initial shoot down, what was going through your mind as the aircraft was hit?

I remember in the airplane we got hit, and simultaneously thinking "We're at 25,000 feet. We're not going to hit the ground right away. We have time to do something and figure out what is going on. If we can't take it all the way back home, we can fly to Syria. It was only like ten miles, so it'd only take us like a minute." This goes back to preparation but not the formal kind. This was more like bar talk of guys, when I was flying the F-4, saying "Flying over Hanoi, that F-4, it's not going to blow up on you. I saw guys screaming ass on fire, but that thing wasn't going to blow up." The F-15E was way better than the F-4 in regards to safety and redundancy. So I knew that we might have gotten hit, but it wasn't like we were in catastrophic danger. I attempted a radio call to let people know we were hit. We lost an airplane two nights before, both guys died, and there wasn't a radio call. Nobody knew what had happened to them at all. I tried.

Were you expecting a search party or rescue?

This is a good one to ask me and Gen Horner about. What he said, and I agree with his decision, is that they shouldn't send a rescue to get me, when I had been shot down and they didn't know what my condition was.

It was a fear of losing additional troops?

Yes, and not knowing if we were ok.

So, it was the fact that you could not radio and tell them you were okay.

Right, they couldn't verify it or anything else. You know it's a high-risk mission. They would have done it. They sent rescue guys in after us, but we were gone. When you are a commander you have to make decisions like that. Sometimes you have to wait and that's what makes it so hard in combat. You have to weigh things out in terms of how many people do I risk to save one person or do I not rescue that one—its not easy. I think that is one of the things that is underestimated—the higher up you go in command, the more decisions like that you're faced with. Do I wish we had been rescued—yes, but I knew based on where we were shot down, based on the assets that could come and get us, it was going to be a long shot. We were so far into Iraq and we knew that before the war. It wasn't a big surprise to me. I wasn't really expecting a rescue–I figured our best bet was to get into Syria and rescue ourselves based on where we were. Just the distance alone—you couldn't come from down from Turkey or up from Saudi Arabia, and fly at night the whole way in and out—those guys were going to be at a lot of risk.

You talked a little bit about the fact that leaders have to make these tough decisions. If you are advising a person who is getting ready going to go into command or a leadership position where they will face these tough decisions, what advice would you give them? How do they prepare for that?

Read a lot of history. Read about other commanders. Air commanders are great, but even other commanders who have had to make tough decisions because that's the only replication you can get. You can't wing it. You have to be prepared before that opportunity comes up. Reading history and understanding what other commanders have had to deal with and how they dealt with it, good and bad, is the only preparation you can do.

Are leaders born or are they developed?

If they were born, I suppose we wouldn't spend all this money on education. I think there's probably a small percentage of people that are natural leaders but everybody can learn leadership and can become better at it. Just because you're not born a "natural leader" doesn't mean you can't be a good leader. Like anything else, it is a skill that can be learned.

Earlier you mentioned that your dad was your biggest influence in your formative years. What leader impacted you the most during your military years? Does anyone stand out?

I'd say Gen Ronald R. Fogleman stands out. He was probably the most authentic leader in terms of what you saw was what you got. He really cared about the mission but he also really cared about the people. Based on my experiences he had some great maxims about being a leader. One was you have three jobs, "Make it happen, make it better, make it last." So when you're in command, you've got to make it happen, whatever the mission is—you've got to do the mission. You've got to make it better because whenever you go into an organization, there's always improvements that can be made because conditions change. And then make it last because you can't just do it on your watch—you've set up processes and systems and train people so that when you're gone they can continue to do the mission you know. Gen Fogleman had thought a lot about it. He had grown up and done everything—to me, he was probably the best leader I saw.

If you had to describe yourself and your experience in one word, what would it be?

Preparation.

You've hit on some common themes here today—preparation, authenticity, toughness. What other leadership attributes stand out to you? What else makes a person a good leader?

You cannot forget the importance of resilience.

Historical accounts describe the Gulf War as the most resounding success in our Nation's history. Do you describe it that way?

As a Nation, as an Air Force, absolutely. Personally, it wasn't a resounding success for me [said jokingly]. Your cadre, your group of people has grown up in a very different Air Force and it's important to keep in mind that the situation and the conflicts we've been in for the last 20 years are not the whole range of conflicts. You could be facing a very different style of conflict in the next five years and if you rely solely on what you learned from your own career, you're going to make some bad mistakes. It's just inevitable because the situation and if the situation is different and you try to apply the same playbook, people are going to die. I want people to understand that they need to think outside of just their own experiences. You can't replicate what you did in the Gulf for the next conflict because you'll possibly be fighting somebody very different and the technology changes. The playbook that John Warden came up with was great for that condition, but it's not necessarily one you can just translate into something else. So I think it was success for the US military and the United States—we achieved our objectives and everything else. I would have preferred not to have been shot down, but you take what you've been given.

Do you consider yourself a hero?

I don't consider myself a hero—I just don't.

Who are your heroes from the Gulf War?

My hero is someone who unfortunately is no longer with us. His name was Captain Bill Andrews. He was an F-16 [Viper] pilot. He won the Air Force Cross for a mission providing close air support to some Army guys and got them out of danger. At the very end of the war, he was on a very similar mission when he got shot down. Andrews got really badly hurt in the shoot down. He broke his leg and a bunch of other stuff during the ejection. He was brought to Baghdad and they treated him pretty badly. He never lost his sense of humor, he never let it get him down. He is the guy you really want to go to war with.

How have your experiences translated to life after the military?

I think I can be pretty resilient and my life perspective has certainly changed.

How do you want to be remembered?

The guy who worked hard and tried to prepare and did the best he could.

What do you think is the most memorable leadership challenge that you had throughout your career?

In 2003, in preparation for the next Gulf War, I had been the Support Group Commander at Incirlik, Turkey for two years, and then I had gone back to the States.

In 2003 we were planning to set up a special operations base and a guy named [Maj Gen] Mike Worden was going to be the wing commander. We knew each other. He called me up and asked me to help set up the base. So, I went over there and it was a challenge from lots of different aspects. One was that we had to spend a lot of time working with Turkish and working with Turkish commanders to get permission for what our guys wanted to do. I was working with people that I never met before, civil engineers guys from Hurlburt AFB, Florida for example and other people so it was a real tag-team effort while trying to earn their trust and get them going in the right direction, keep them motivated and working with a bunch of a communications people that I didn't know, along with cops and everything else. Eventually though, although we set things up, they were never used because the parliament voted against entry into the war and we had to leave. Those couple months were still the most challenging times from a leadership perspective.

Looking back on your experiences, is there anything that still keeps you up at night?

Sure, I mean doesn't really keep me up now so much but it has in the past. That is, second guessing things that happened during the mission and when I was a POW. Did I hold out as much as I could? Did I serve the right way or was I not good enough? Because no one is going to tell you that you screwed up or you did good. That bothered me a lot before and although it doesn't bother me as much today that would be the one thing that used to keep me up.

With the benefit of hindsight, what do you wish you had done differently during that time?

I had a bad feeling about what we were trying to do by getting close to the building when we were making our way to Syria, but at the time we'd been sleeping out in the open for three days. We had not had any food and there had been so many other plans where I said, "No, that's not a good idea." I think at the time, I was just like, "Well, yeah, maybe this is not the worst idea in the world," but I had a feeling in the back of my mind that said, "This is not a good idea."

What do you think about post-Gulf War changes? Does anything stand out?

I think the move to precision is good. Stealth is definitely the way to go. My biggest concern is if people are just fixating on DESERT STORM, they need to think more broadly about war. You can think of scenarios where we will go against an adversary that is not like the Taliban. Since the first 48 hours of DESERT STORM, we haven't had to contest air superiority. That's fortunate in one way, but unfortunate in others because people don't understand how important it is.

How did your experiences affect your family or your work-life balance?

It was hard on them and it continues to be hard on them today. One of my daughters turned three when I was shot down. She had been just like a normal three-year-old that was talking and then the whole time I was shot down she wouldn't speak. So yes, it took a big toll on them. There's really no way to make up for that.

What advice would you give to the families of leaders who are about to be put in these difficult positions?

For the military leader going into these positions, it would pay dividends to be honest with your family members. There is no sense in sugar coating it. You know you have to make clear to them that your business is warfare and at some point you're going to have to go to war. You're going to have to deploy, you're going to have to go into harm's way. If you never have to do that, then nothing is wasted but I think the family needs to be as prepared as the service member for what could possibly happen and you know I speak about this not just from my own experience but experiences I've heard about from other people. So, yeah I think it's important for families to be prepared too, not in a survivalist kind of way, but just understanding what it means to be in the military and what that's going to entail for them. Deploying may be an old hat for people nowadays, but maybe it won't be in 10 years. I don't know.

Do you think you were a good leader?

I think I had some good leadership characteristics.

What would you say you did well?

When I was a group commander I think I did a pretty good job mentoring my squadron commanders and I tried to set an example for how leaders should be, especially in terms of not sitting in the office and getting to know everything about the base and what was going on and all the people working there. I tried to motivate everyone towards a common goal.

What do you think you could have done a little better?

I could have been more patient with some people. I think I judged some people too quickly. Gen Fogleman had a great saying, he said, "You have to use every tool in the toolbox. Some people are good at some things and not as good at other things. You need to make sure you're using the right person for the right job, and sometimes you may have to train the person to do the job if they're not doing it right."

Speaking of leadership challenges, would you have wanted General Horner's job?

I would have loved Gen Horner's job if I could have gotten promoted to that level! [said jokingly] I do have an appreciation for the burden he was faced with. Being a commander is not about the parking spot and the nice office. That kind of responsibility has life and death implications and you have to be prepared for that.

What advice would you give to a young 16- or 17-year old who says they want to join the military?

They should want to serve their country, but it won't always be under pleasant conditions. They need to accept that. I'm not sure a 17-year old can really understand that, but that's what I would tell them. I'm not sure many 22-year olds would understand that.

What about guys at ACSC, getting ready for command?

Keep learning, keep preparing. You can always learn more about the profession of arms. You can always learn more about leadership. You can't depend on one year at ACSC either. You must have a serious self-education program.

Are there any books on leaderships that stand out to you?

One that a lot of people might not think about that's not as much about leadership as much as it is preparing for leadership, and that is *The Patton Papers*. These are the papers and letters written largely by GEN George Patton. He was a person who spent a lot of time preparing for war, and he did that when he was a junior officer. I'm not recommending anybody copy George's style, but if you read those books you get a good sense of how much effort he put into being prepared. He obviously grew up in a different military. The interwar years were very different than today in terms of lifestyle. He didn't have distractions like YouTube, Facebook, and Instagram, but leaders can still carve some time away from that to read a little and prepare.

You've had some incredible experiences. What are your final thoughts? What

should people take from your story?

I think that people and history have proven that time after time, they can overcome a lot of obstacles. A lot of it boils down to will and preparation. That is true for things in their personal lives and their professional lives. It's easy to dismiss that because it sounds trite and cliché. But it's cliché for a reason; it's true.

As a young kid, Col Thomas Griffith never imagined that one day he would be considered a hero—in fact, he still does not consider himself a hero. Yet, at the age of 34, his name was being mentioned by almost every news outlet in the world. He chose to serve his country, and in doing so, he paid a tremendous price. He is a hero and a national treasure. Before interviewing Col Griffith, we knew we would be interviewing a war hero. After interviewing him, we learned that not only is he a hero, he is great human being. Col Griffith is the most humble and kind person one can hope to meet. It was extremely humbling to interview this great American. While his story has impacted thousands, it is his humility, his sense of humor and his willingness to give back, that was the most striking. His will, preparation, and strength serve as an example for future leaders to follow.

■ Interviewed by Maj Mike Ryan, US Air Force, and Maj Christopher Jefferson, US Air Force.

Part II:

Forging Leaders

LESSONS IN LEADERSHIP

Each chapter of Part I: Honoring Legacies, stands alone as an individual piece of history. The focus of Part I remains on the personal accounts of the Eagles in order to honor their legacies and allow future generations to read the history of DESERT STORM through the words of those who were there. The focus of Part II: Forging Leaders, is placed on the leadership lessons each team took from meeting and interviewing their Eagles. Teams also provide a personal account of their interview to allow the reader greater insight into each Eagle. While this section represents the personal opinion of the authors, the intent is to carry the leadership thread throughout this work and demonstrate the process of learning what it means to lead from those who have come before us. Gen Horner, Maj Gen Feest, Maj Gen Johnson, CWO Hunter, Gen Kross, Brig Gen Cornum, Col Warden, and Col Griffith, the following is what we have learned from you.

GOE Students (L-R). Front Row: Maj Vanessa Wilcox, Maj Erika Chute, Maj David Underwood, Maj Jessica Guarini, Maj Adrienne Stahl, and Maj Michael Ryan. Middle Row: Maj Shannon O'Neal, Maj Dustin Wade, Maj Heather Demis, Maj Fred Merrill III, and LCDR R. Thomas Griffin. Back Row: MAJ Thomas Bramanti, Maj Will Powell, Maj Christopher Jefferson, Chaplain (Maj) Christian Williams, and Maj Stanley Grodrian.

Charles A. "Chuck" Horner

Don't Take the Easy Way Out

By MAJ Thomas Bramanti and Maj Vanessa Wilcox

So there were were...(because that is how all great flyer's stories start, and this is a story about a great flyer) pulling into Gen Horner's driveway. I hate to admit it, but we were nervous. How often is it that you get to interview a retired four-star general, let alone drop by his house for the interview? In addition, this particular general has an extensive amount of material out there about him and his career. From multiple interviews to a Tom Clancy book, essentially anything we wanted to know about him was at our disposal. How were we to ask him the appropriate questions without sounding like majors who had not done their homework? How did we walk into this house and find that tidbit of information that would help shape us, and the rest of our ACSC class, as future leaders?

Our nerves subsided as we entered the house and could see the ocean through the trees out the back windows of the house. Gen Horner was eager to get down to business and start the interview. With all the standard questions, Gen Horner provided the same level of consistency in his answers that we had seen in his previous interviews. While the Gulf War was the highlight of his career and the focus of our chapter, a personal goal was to learn about leadership from someone who had led so many. There were practical leadership lessons like, "your NCOs will provide you a good sense of how your organization is doing" and "the importance of building and maintaining relationships." As well as others that were refreshing to hear from a senior officer such as, "don't ever solely focus on making the next rank at the expense of the mission." Finally, he told us that he believed the most important aspects of leadership were "selflessness" and "having the courage necessary to make a difficult decision."

While his words on leadership were valuable lessons for us, through listening to his stories and spending time studying Gen Horner, the most significant lesson in leadership we took from this experience was "don't take the easy way out." This theme not only resonated in his life stories, but we also witnessed it through our personal interaction with him during the Gathering of Eagles program. Whether you read Gen Horner's words in this book or a briefing he gave to the President of the United States on the DESERT STORM air campaign plan, it is evident that Gen Horner only allows himself to give his absolute best in all he does, despite the personal cost. To leave behind a legacy that not only you but also future generations would be proud of, it is sometimes necessary to take the route of most resistance. This path requires both selflessness and an extraordinary degree of courage. If you need inspiration, we leave you with Gen Horner's words, "You can kill me, but you can't eat me."

Gregory Feest

Be the Best at Whatever Job You Have and Don't Worry About Anything Else

By Maj David Underwood and Maj Will Powell

Gen Feest cemented his place in history as the first pilot to drop bombs in both Operations JUST CAUSE and DESERT STORM, but his impact on the Air Force was much greater than those remarkable events. From training the next generation of aviators as a squadron, group, and numbered air force commander; to leading the largest, most diverse expeditionary wing in the Air Force supporting combat operations in Afghanistan and Iraq; to directing the Air Force's entire safety program; Gen Feest ensured and preserved the combat readiness of our Air Force. He has a passion for leading, training, and preparing others to help them achieve excellence.

We could sense this passion as we interviewed Gen Feest in his home over the course of two days. He is quick to highlight the contributions of others and how lucky he was for his leadership to choose him to lead the first attacks in both operations. Perhaps luck played a part, as he contends, but he truly excelled in everything he did. His selection as lead pilot in two operations and his later appointment to multiple command positions demonstrate how opportunities arise when individuals strive always to do their best.

This is the lesson we took from Gen Feest; "Be the best at whatever job you have and don't worry about anything else." Even if the job you currently have is not what you truly want, the job may be preparing you for a future role in ways you do not yet understand or cannot predict. Gen Feest did not want to be a navigator initially, but he excelled at it. He earned a pilot slot as a result of his performance in the aircraft, which led him to the F-15C, the F-117A, and then as the lead pilot on the opening night of one of the most successful air campaigns in history.

Gen Feest had an incredible career in the Air Force. He absorbed everything he could from leading thousands of people in multiple command positions and is eager to share his lessons learned. It was a joy and an honor to interview Gen Feest and to hear his thoughts on living and leading.

Paul T. "PJ" Johnson

Be Humble, Be A Coach, and Don't Forget to Have Fun

By Maj Erika Chute and Maj Jessica Guarini

Crunch! All of a sudden we were sandwiched in between two vehicles on I-65. Our hopes to interview Maj Gen "PJ" Johnson had been dashed for the day. When we called to let him know about the mishap, his voice was calming and understanding. We would try again another day. A week later, we attempted the interview again. As we navigated through the quaint town of Henderson, Tennessee, we noted the peaceful and serene surroundings. What a change of pace this lifestyle must be for a man who just recently retired from such a fast-paced career. Our initial impressions were further affirmed when we pulled up to a house in a sleepy wooded community.

As we approached his door, a sense of mild trepidation came over us. We were about to meet an airpower legend. Our anxious thoughts were quickly quelled as Gen Johnson warmly welcomed us inside. Filled with family pictures and mementos, the beautifully decorated home was neat and orderly. It was apparent faith played a significant role in his life. He asked if his wife, Patricia, could join us for the interview and expressed that it was her birthday. We did not take the Johnson's sacrifice of time lightly and were thankful they allowed us to be part of this special day. Shortly after beginning the interview and asking our initial questions, we were transported back in time. As Gen Johnson described his experiences from working on a hog farm to transitioning to flying a "hog" (A-10), we felt as if we were a part of his journey. Expressing sentiment for each person who influenced him along the way, Gen Johnson's detailed memory impressed us. He had a way of describing each person and expressing how important each one was to the success of his career. Launching into his experience in DESERT STORM, the tone changed slightly. He became focused and intent on communicating his unit's objectives and described the readiness of the force. You could tell in his eyes and his voice that both the search and rescue mission, and the day that his aircraft, 2664, came home on a prayer, dramatically shaped the remainder of his career.

Gen Johnson helped us understand the importance of being confident in your decisions, including your family in your choices, and always remembering to have fun. In his words, "Enjoy what you're doing and the job you have. Stick around, and you'll do a different one." He is living evidence that the qualities necessary for success are more than just a list; character remains the bedrock of leadership. He epitomizes the trinity of balance in the spiritual, physical, and mental aspects of life. Perhaps the most compelling testimony to Gen Johnson's character, his marriage of 37 years, demonstrates that the wingman concept transcends military culture into one's private life.

Before we departed, we noticed a portrait of a man in the cockpit of an A-10 kissing a woman. Patricia relayed that a photo of her and PJ, inspired her daughter-in-law to paint the keepsake. As she described the history of the photo, the teamwork between this amazing couple was further validated.

On the drive back, we reflected on the day's experience. We were humbled to have had the opportunity to interact with such a great leader and to catch a glimpse into his life and family. Although the interview only lasted a few hours, the leadership lessons captured will resonate with us for a lifetime.

Guy Hunter, Jr.

Be the Expert, Be Consistent, and Remember to Laugh

By LCDR R. Thomas Griffin and Maj Fred "Trey" Merrill III

From the moment we arrived at the Hunters' beautiful beach home, a cool breeze from the ocean greeted us, providing an immediate sense of serenity and peace. After hearing Guy Hunter relay his extraordinary experience of being held captive during DESERT STORM, we found ourselves comparing the stark difference between his captivity with where we were sitting. The disparity was even more apparent as the Hunters guided us out to their walkway leading directly to the beach. We were clearly in a place where the horrors of war dare not go.

Meeting Guy Hunter was a unique and enjoyable experience. With everything he went through, he was always quick to joke and laugh. There is an old saying, "when you laugh, you take a mini vacation." Guy Hunter takes a lot of vacations! Not only did he greet us with amazing warmth, he was very curious to know what we were studying. Both Mary and Guy wanted to know about us: our lives and our families. It was incredibly gracious of the Hunters to ask about us when we were there to hear their stories.

You could tell that family was an important part of their lives. Photos of children and grandchildren graced the walls, including a beautiful wedding photo of the Hunters. Guy wore Marine Corps dress blues, and Mary wore an elegant wedding dress. They appeared larger than life: Guy the military hero and Mary resembling a beautiful Hollywood actress.

The Hunters made us feel like we were family, which made the interview a smooth process. It afforded us the leeway to veer outside our structured questions, and explore the areas of Guy's life, and experience, which were not readily apparent.

While interviewing Guy, two leadership lessons became apparent: 1) be an expert and 2) be consistent. He was incredibly industrious in his knowledge and an expert in his profession. Numerous times, Guy commented on the importance of knowing "what the hell you are supposed to be doing" in a job. He explained that having expertise transcended rank; it was important for leaders and followers alike. However, Guy's advice to those specifically in leadership roles was to be consistent. As a leader in the military, there is a lot to think about and consider. Consistency, he explained, will see you through.

While in his interview, Guy commented that leadership was simple, he exhibited numerous leadership qualities that seemed to come naturally to him. He is a man of honesty and integrity. He displayed courage and honor while in captivity, and despite his trying circumstances, found opportunities to make jokes and laugh. Nothing was going to crush his spirit.

Perhaps that optimism and ability to endure dark days is something that we all can learn from CWO Hunter. 26 years ago, Guy sat in a cell cleaning sores from his body and contemplating a torture session where he might lose his fingers. Today, he is living his version of paradise, on the beach, surrounded by family. Though most of us will never have to endure dark times to the degree that Guy faced, his story reminds us that if we can find a way to get through hard times, somehow finding an opportunity to laugh, better days and blue skies await.

Walter Kross
Embrace the Challenge
By Maj Heather Demis and CH (Maj) Christian Williams

"Challenges are what make life interesting; overcoming them is what makes life meaningful."

JOSHUA J. MARINE

Many thoughts were running through our head as we arrived at Gen Kross' beautiful home. We felt honored to meet with him, but at the same time felt the pressure of capturing the thoughts and experiences of this great leader in a single interview. From the moment we entered his conference room, it was evident that he continued to live a life of service before self. On the table, were bottles of water and cookies, freshly baked by his wonderful wife, Kay. We were captured by the humility of this DESERT STORM icon and the graciousness of his family in welcoming us into their home. It also became apparent that relationships are important to Gen Kross as he spent time asking about our lives and careers.

During the interview, we were impressed by his ability to recall events during DESERT STORM, which dramatically elevated our academic understanding of the conflict and of what it meant to lead during that time. President John F. Kennedy said, "Ask not what your country can do for you, but what you can do for your country." Gen Kross is the epitome this quote. With the many constraints and limitations encountered in planning and executing DESERT STORM, he never complained or gave excuses. He and his team met the challenges head-on and our Nation and military gained from their success.

A part of Gen Kross' life that was especially inspiring was his ability to balance a distinguished career and time with his family. Photos of his daughter and grandchild graced the walls, including a beautiful picture of Gen Kross and his wife. It was obvious he believed the support and sacrifice of his family played a significant role in his success. Although family separations and long hours at work were difficult, it was evident that Mrs. Kross played a pivotal role in helping their family navigate and overcome life's challenges.

The privilege of meeting Gen Kross was a remarkable experience. He imparted life lessons crucial to the development of any leader throughout the interview. During our time with him, two leadership lessons became very apparent: 1) No one knows themselves until they are truly challenged; and 2) No organization knows about itself until it is stressed. It is during these challenges or stressful times in life that a person's character is forged. In facing these times, one discovers what lies inside them and decides what lies before them. How a person reacts to challenges is a mark of their true character. We are incredibly grateful to have met Gen Kross and his family, and for the leadership lessons we took from this experience. We feel honored to continue his legacy in service to our Nation and for the opportunity to share his story.

Rhonda L. Cornum

You are Stronger than You Think

By Maj Stanley Grodrian and Maj Shannon O'Neal

As we drove through the rolling hills of the horse capital of the world, we were in awe of the antebellum architecture and huge horse farms surrounding us. Neither of us knew what to expect as we drove the long winding driveway that led to Gen Cornum's house but we both expected something different than what we found. We pulled up to a beautiful, small stone house, originally built in the early 1800s, sitting in the middle of pastureland. Gen Cornum was at the door, greeting us before we got out of the vehicle. The inside was rustic and comfortable with a modern kitchen and large open sitting room. Gen Cornum welcomed us into her home with warmth and a comfortable, easy manner. Our conversation was often punctuated by laughter and one could not help but feel the positive energy emanating from our host. With everything that Gen Cornum has endured, some might find her positivity curious, but you do not have to spend much time with her to see that this positive outlook on life is the source of her strength and resilience.

In addition to being one of only three survivors out of an eight-person helicopter crew that was shot down during DESERT STORM, and subsequently enduring time as a POW, Gen Cornum survived breast cancer and a traumatic accident where her vehicle with attached horse trailer rolled off an embankment. Her first thought after the rollover incident was gratitude; the horses were safe, she had not brought one of her prized dogs, and she was not severely injured. This positive outlook on life and the challenges we all must face is what makes her more than a survivor. To survive has the connotation of merely enduring something, but Gen Cornum did not merely survive events that would threaten to destroy even the toughest among us; she crushed them and flourished. When asked if she had any regrets or if she would do anything differently, she responded with a simple "no, I'm good." We were struck with the realization that this might be the toughest person we had ever met. If a leader is to understand resiliency, Gen Cornum is the example.

As Gen Cornum's heroic story of survival and resilience as a POW during DESERT STORM is well documented in the book, She Went to War, the questions we chose to ask during the interview focused on leadership and life lessons we could learn from her enigmatic and distinguished career. When we asked a question regarding her being a woman and a military officer, we quickly learned that this particular question was neither appropriate nor relevant to Gen Cornum. During her time in the Army, the fact that she was a woman was not a thought in her mind. She was simply a human being, a military officer, and a leader.

Gen Cornum's lessons on leadership are straightforward and simple; trust your people and lead them in the manner that is best for the mission and individual. Each person is different, but that does not mean to pander to each. Treat all people with respect and due diligence and the rest will fall into place. It does not matter the orientation, religion, or gender of an individual in your command. When asked how to make hard decisions, her answer was straightforward and direct. Make an informed decision based on the information you have, but make a decision. Making a decision is the easiest thing to do because it has to be done. Take out the emotion, be honest with yourself, uncover the facts, and make the decision. Her leadership style left us thinking about the adage: "lead, follow, or get the hell out of the way."

To say that Gen Cornum is an officer we would willingly emulate is an understatement. She continually sets the example in both her personal and professional life. She is intelligent, strong both mentally and psychically, and unapologetic. Despite her high-profile life, she places no emphasis on fame. Current and future generations can glean important lessons from studying Gen Cornum's example and words of wisdom. We are deeply grateful to Gen Cornum for opening her home to us and for allowing us the honor ofmeeting such an extraordinary leader and hero. She reminds us all that we are stronger than we think.

John A. Warden III
Be Bold, Be Persistent, Do the Right Thing

By Maj Adrienne Stahl and Maj Dustin Wade

Upon arriving to meet and interview Col John A. Warden III, we were ushered to a smart, tidy, but not overly elaborate, conference room. Col Warden entered the room with a welcoming manner and just a touch of old-school formality. Our conversation started right away, and his answers were very thorough and direct, of a cerebral nature we might have expected from a man known as a new-age airpower theorist. It was obvious he had a mind built for strategy. He strung together extraordinarily complex stories, articulating the finest details of events that happened 25 years ago by building a background, providing context, and guiding us through his thought processes. Throughout the interview, he displayed a seemingly absolute objectivity, thoughtfully considering all viewpoints when recalling challenging scenarios. He was rational and logical. His matter-of-fact candor and viewpoints were unbridled by emotion and delivered with an air of calm and confidence, which seemed central to his essence as a leader.

It became apparent when listening to his stories that even when up against resistance to change he proposed or ideas he had, he never compromised his beliefs, would not filter his thoughts for another's benefit, and did not shy away from confrontation or conflict if he thought a wrong needed to be made right. He demonstrated a steadfast consistency in beliefs, principles, and values—to always do and say what he felt was right for the greater good, with a complete disregard for personal gain. He inherently understood these attributes to be necessary for moving an organization forward, especially when dramatic change is required. He imparted to us that this determination and mold-breaking attitude will likely meet, and will potentially cause, heavy institutional and personal resistance. To be able to drive change successfully requires boldness and persistence, as well as a certain level of empathy and style. He warned that if you do not manage resistance correctly, the change will likely fail or could have career-impacting implications. Col Warden's thoughts on "open-planning" and how to "lead up" effectively can help to mitigate these challenges and improve your chance of success. At the end the day, however, it is vital to decide which ideas are worth fighting for because you may have to "fall on your sword" to see them through.

From this interview, we also learned that groundbreaking ideas do not usually come from an instant epiphany. Instead, ideas are grown and cultivated over extended periods from experiences, questioning of why things are a certain way, and asking if there is a better way to accomplish the task. Col Warden's concepts on the role of airpower were developed over time, through his combat experiences and exposure to the strategic level of Air Force thought and function. His thoughts on airpower were shaped as he continually questioned the role and effectiveness of airpower. Col Warden's concepts represent a step forward in airpower theory from the stagnation of the Cold War era. He knew, and taught many, that there would no longer be one enemy and one option for war, and he helped prove that airpower was capable of more than many believed possible.

These leadership attributes—boldness, candor, persistence, doing the right thing with disregard for personal gain, and a habit of questioning the status quo—resulted in the development and execution of the most successful air campaign seen to date, where airpower was used in ways the world had never seen.

Thomas Griffith

Will and Preparation

By Maj Christopher Jefferson and Maj Michael Ryan

Col Griffith's story is one of courage, resilience, and preparation. He deployed as a weapons system operator in the F-15E as DESERT STORM began, and he was shot down on the fourth night of the war. He ejected at 25,000 feet, and after safely landing on the ground, Maj Griffith was able to avoid detection and capture for two days. He was ultimately captured and became a POW, which is an ordeal that most cannot imagine. His experience involved uncertainty of what was to come, contending with enemy interrogation tactics, and dealing with the profound impact of this event on his family. Colonel Griffith does not consider himself a hero, but his humility is befitting this man who has lived a life of service to country. He fought wars on behalf of the Nation and in doing so, paid a remarkable price. He also chose to serve a full career in the military, even after his time as POW. While Col Griffith will not call himself a hero, he is just that—his acts of service and sacrifice make him a hero and a national treasure.

After hearing his story and understanding the details of his experience, it is important to pass on lessons that can be used to forge leaders of future generations. The critical question for those who follow is: "how can a person survive undue hardship and adversity? Col Griffith provides two crucial leadership characteristics to help answer this question—will and preparation.

A future leader must have the will to achieve mission success. Sheer will helped Col Griffith deal with the uncertainty of his experience. He was interrogated, blindfolded, thrown into the trunks of cars, and forced into solitary confinement with only one meal a day. Moreover, he dealt with the mental anguish of not knowing if he would ever see his young children or his family again. Through this time, he maintained his will to serve and to survive. Col Griffith's experience demonstrates the importance of resolve and resilience in the face of hardship. In addition to maintaining the will to survive and succeed, the most resounding lesson one can learn from Col Griffith is the importance of preparation. Throughout his interview, he talked about his father's teachings as a young child, his preparation as a cadet at the Air Force Academy, and his training leading up to DESERT STORM. He also emphasized the importance of reading history and learning lessons from other commanders. From talking to him, it became apparent that preparation and training were his most valued attributes. When he faced his darkest hour and his greatest degree of uncertainty, he resorted back to his training, and that is what got him through. These are lessons applicable to all leaders as adversity is certain, though not often to the degree that Col Griffith faced. Col Griffith's experiences demonstrate that preparation and resolve can get a person through life's toughest challenges.

Col Griffith is a great American and an even better person. It is an honor to don the same uniform as him. He paid a tremendous price in service to this great Nation, and we honor his sacrifice. It is our sincerest hope that future generations will learn from these lessons and insights and from reading Col Griffith's words.

TAKE AWAYS

Don't Take the Easy Way Out

Be the Best at Whatever Job You Have and Don't Worry About Anything Else

Be Humble, Be a Coach, and Don't Forget to Have Fun

Be the Expert, Be Consistent, and Remember to Laugh

Embrace the Challenge

You are Stronger than You Think

Be Bold, Be Persistent, Do the Right Thing

Will and Preparation

These are the lessons we will take with us from this phenomenal experience. General Horner, Major General Feest, Major General Johnson, Chief Warrant Officer Hunter, General Kross, Brigadier General Cornum, Colonel Warden, and Colonel Griffith, thank you for the example of what it means to live a life of service and for the sacrifice that requires each day. We are humbled and honored to share your words with those who follow in your footsteps. To you, and to all those airpower pioneers and legends before you who share your distinguished title of Eagle, may your legacies live on in those brave young men and women who follow in answering the call of a grateful Nation.

CHRONOLOGY

AUGUST 1990

2 Aug	Iraq invades Kuwait.
4 Aug	Gen Hansford Johnson, Commander in Chief, TRANSCOM, activates his CAT.
5 Aug	Horner briefs President George H.W. Bush on air operations. Bush declares invasion "will not stand."
6 Aug	King Fahd meets with Secretary of Defense Cheney, requests US military assistance. Horner arrives in Saudi Arabia as CENTCOM commander forward.
7 Aug	C-Day—beginning of deployment—DESERT SHIELD begins.
8 Aug	Initial US Air Force fighter planes arrive in Saudi Arabia. GEN Schwarzkopf calls Gen Loh, at the Pentagon to ask for assistance in developing the Air Campaign Plan. Col Warden tasked to provide it.
10 Aug	Warden meets with Schwarzkopf in Tampa to outline proposed air campaign, INSTANT THUNDER.
11 Aug	Air campaign plan briefed to GEN General Powell.
20 Aug	Air campaign plan briefed to Horner.

SEPTEMBER 1990

18 Sep	Schwarzkopf asks Army planners to begin work on ground offensive.

OCTOBER 1990

10 Oct	CENTCOM's One Corps Concept unveiled at White House.
21 Oct	Powell flies to Riyadh to discuss offensive plans.
31 Oct	Bush decides to double US forces in Saudi Arabia; decision kept secret until 8 November.

NOVEMBER 1990

29 Nov	UN Security Council authorizes use of "all means necessary" to eject Iraq from Kuwait.

DECEMBER 1990

6 Dec	First ship carrying VII Corps equipment arrives in Saudi Arabia from Germany.

JANUARY 1991

9 Jan	SECSTATE Baker meets Tariq Aziz in Geneva in unsuccessful effort to find a peaceful solution.
12 Jan	Congress authorizes use of force.
15 Jan	UN deadline for Iraqi withdrawal.
17 Jan	Allied attack begins at 0238 with Apache strike At 0251 Gen Feest leads the strike package of F-117As and releases the first GBU-27s in Baghdad targeting key nodes of Iraq's SAM network.
18 Jan	First Scuds hit Israel. First American air attacks are launched from Turkey. Hunter and Cliff Acree are shot down.
19 Jan	David Eberly and Griffith are shot down.
20 Jan	Lawrence Eagleburger and Paul Wolfowitz arrive in Tel Aviv.
23 Jan	Intense attack against Iraqi aircraft shelters begins.
24 Jan	PJ Johnson's A-10 aircraft is hit by enemy fire.
26 Jan	F-111s attack oil manifolds at Al Ahmadi in effort to counter Iraqi sabotage.

29 Jan	Iraqis attack Khafji and other border positions. Allied pilots begin flying combat air patrols to thwart Iraqi flights to Iran.
30 Jan	Cheney dispatches Delta Force to Saudi Arabia to hunt for Scuds.
31 Jan	Khafji is recaptured. Some POWs moved to the "Biltmore" prison.

FEBRUARY 1991

1 Feb	Last Tomahawk missiles are launched in attack on Baghdad airfield.
2 Feb	Schwarzkopf formally decides against amphibious landing in Kuwait. February 5 First "tank-plinking" mission flown.
6 Feb	VII Corps finishes closing in theater with arrival of final 3rd Armored Division equipment.
8 Feb	Cheney and Powell fly to Riyadh for final review of ground war plans. Johnson locates and leads a successful CSAR "SANDY 57" mission of a downed Navy F-14 Tomcat pilot deep behind enemy lines.
11 Feb	Al Firdos bunker in Baghdad suburb added to master attack plan.
13 Feb	Strike on Al Firdos bunker kills more than two hundred civilians and leads to restrictions on strategic bombing campaign.
15 Feb	Radio Baghdad suggests Iraq willing to withdraw, Bush rejects proposal as a "cruel hoax."
16 Feb	VII Corps moves into final attack positions.
21 Feb	CIA and Pentagon officials meet at White House to air differences over battledamage assessment. Bush sets deadline of 1200 23 Feb, for Iraqi withdrawal.
23 Feb	Stealth fighters attack Iraqi intelligence headquarters, unaware that allied POWs are inside. Army Special Forces teams inserted deep into Iraq.
24 Feb	Ground attack begins. Schwarzkopf decides to accelerate main attack of VII Corps by fifteen hours. F-117A bombs prison facility in Baghdad where Hunter and Acree were held.
25 Feb	101st Airborne Division cuts Highway 8 in Euphrates Valley. Iraqis counterattack 1st Marine Division. Scud destroys barracks in Al Khobar, killing 28 Americans and wounding 98.
26 Feb	Iraqis flee Kuwait City. VII Corps hits Republican Guard in Battle of 73 Easting.
27 Feb	Maj Cornum is shot down and taken captive.
28 Feb	Cease-fire takes effect at 0800.

MARCH 1991

2 Mar	24th Infantry Division fights Hammurabi Division as it flees; destroys 600 vehicles.
3 Mar	Schwarzkopf meets Iraqi generals at Safwan.
5 Mar	Griffith, Cornum, and most other POWs are released.
6 Mar	Hunter, Acree, and Eberly are released.
10 Mar	R-Day beginning of redeployment.
14 Mar	Desert Express discontinued.

JUNE 1991

8 Jun	Victory parade in Washington D.C

*Dates and hours are in Riyadh time.

ACRONYMS

AAA .... Anti-Aircraft Artillery
AB .... Air Base
ACC .... Air Combat Command
ACSC .... Air Command and Staff College
ADF............................... Automatic Direction Finder
AETC .... Air Education and Training Command
AFB .... Air Force Base
AFOQT.......................... Air Force Officer Qualifying Test
AFSOC.......................... Air Force Special Operations Command
ALCE .... Airlift Control Element
AMC .... Air Mobility Command
ANGLICO .... Air Naval Gunfire Liaison Company
AOR .... Area of Responsibility
APOD .... Aerial Port of Debarkation
ASTRA......................... Air Staff Training Assignment
ATO .... Air Tasking Order
AWACS .... Airborne Warning and Control System
AWOL........................... Absent Without Leave
C3.................................. Command, Control, and and Communications
CAOC .... Combined Air Operations Center
CAP............................... Combat Air Patrol
CAS .... Close Air Suppor
CAT............................... Crisis Action Team
CC .... Commander
CENTAF .... United States Air Force Central Command
CENTCOM .... Central Command
CENTO .... Central Treaty Organization
CIA................................ Central Intelligence Agency
CINC .... Commander in Chief
CINCEUR.................... Commander in Chief, European Command
CO................................. Commanding Officer
CONUS .... Continental United States
CRAF .... Civil Reserve Air Fleet
C-RATS........................ Combat Rations
CSAR .... Combat Search and Rescue
CSH.............................. Combat Support Hospital
CWO .... Chief Warrant Officer
DEAD .... Destruction of Enemy Air Defenses
DG .... Distinguished Graduate
DIA .... Defense Intelligence Agency
DMPI .... Desired Mean Point of Impact
DO .... Director of Operations
DOD .... Department of Defense
DOT............................. Department of Transportation
ECS .... Environmental Control System
EUCOM .... European Command
EWO .... Electronic Warfare Officer
FAA .... Federal Aviation Administration
FAC .... Forward Air Control
FOL.............................. Forward Operating Location

FSS .... Fast Sealift Ship
GATM .... Global Air Traffic Management
GBU .... Guided Bomb Unit
GCI .... Ground Control Intercept
GDSS .... Global Decision Support System
GOE .... Gathering of Eagles
GPS .... Global Positioning System
GTN .... Global Transportation Network
HAZMAT .... Hazardous Material
HF .... High Frequency
INS .... Inertial Navigation System
IOC..............................Initial Operational Capability, Intercept Operations Center
ISOPREP .... Isolated Personnel Report
IT .... Information Technology
ITV .... In-Transit Visibility
J3 .... Joint Operations Officer
J4 .... Joint Logistics Officer
JATO .... Jet-Assisted Takeoff
JCS .... Joint Chief of Staff
JDAM .... Joint Direct Attack Munition
JFACC .... Joint Forces Air Component Commander
JMEM........................... Joint Munitions Effectiveness Manual
JOPES .... Joint Operation Planning and Execution Systems
JSTARS......................... Joint Surveillance Target Attack Radar System
KIA .... Killed in Action
LNO .... Liaison Officer
LORAN........................ Long-Range Navigation
LTG .... Lieutenant General
MAC............................. Military Airlift Command
MARAD....................... Maritime Administration
MEF .... Marine Expeditionary Force
MHE .... Material Handling Equipment
MOS............................. Military Occupational Specialty
MSC .... Military Sealift Command
MTBF........................... Mean Time Between Failure
MTMC .... Military Traffic Management Command
NASA .... National Aeronautics and Space Administration
NATO........................... North Atlantic Treaty Organization
NCO .... Non-Commissioned Officer
NDRF........................... National Defense Reserve Fleet
NRO .... National Reconnaissance Office
NSA.............................. National Security Agency
OCA .... Offensive Counter Air
OCS.............................. Officer Candidate School
OER .... Officer Evaluation Report
OPLAN .... Operational Plan
OPSO .... Operations Officer
ORI............................... Operational Readiness Inspections
OTS.............................. Officer Training School
PGM .... Precision Guided Munitions
POL .... Petroleum, Oil, and Lubricant
POW .... Prisoner of War

PR .... Public Relations
PRC .... Portable Radio Communication
RO/RO .... Roll-On/Roll-Off
ROTC........................... Reserve Officer Training Corps
RRF .... Ready Reserve Force
RWR............................. Radar Warning Receiver
SAASS .... School of Advanced Air and Space Studies
SAC .... Strategic Air Command
SAM .... Surface-to-Air Missile
SCUD........................... Subsonic Cruise Unarmed Decoy
SECDEF .... Secretary of Defense
SECSTATE .... Secretary of State
SEFE .... Standards and Evaluations Flight Examiner
SERE .... Survival Evasion Resistance and Escape
SNCO.......................... Senior Non-Commissioned Officer
SOS .... Squadron Officer School
SPACECOM .... Space Command
SPOD .... Seaports of Debarkation
TAC .... Tactical Air Command
TACAN .... Tactical Air Navigation System
TCC............................. Transportation Component Commands
TDY .... Temporary Duty
TFR .... Terrain-Following Radar
TFI .... Total Force Integration
TFS.............................. Tactical Fighter Squadron
TPFDD .... Time Phased Force Deployment Data
TRANSCOM .... Transportation Command
TTPs............................ Tactics, Techniques, and Procedures
ULN .... Unit Line Number
USA .... United States Army
USAF........................... United States Air Force
USAFE .... United States Air Forces Europe
USCINCTRANS....Commander-in-Chief, United States Transportation Command
USMC .... United States Marine Corps
USN............................. United States Navy
VA................................ Veterans Affairs
VIP .... Very Important Person
VMO .... Marine Observation Squadron
WMD .... Weapons of Mass Destruction
WSO............................ Weapons System Officer
WWI .... World War I
WWII .... World War II
WWMCCS.................. Worldwide Military Command and Control
XO .... Executive Officer
XOX............................. Directorate for Warfighting Concepts
XOXW......................... Directorate for Warfighting Concepts, Concepts Division

LIST OF EAGLES: 1982-2017

ARGENTINA

Horacio Mir Gonzales .... FAA
Augusto Bedacarratz................... ARA

AUSTRALIA

Andrew W. Barr .... RAAF
Clive R. Caldwell....................... RAAF
Robert H. M. Gibbes .... RAAF

BANGLADESH

Saiful Azam .... BAF

BELGIUM

Raymond A. Lallemant............... RAF

CANADA

Russell Bannock .... RCAF
Leonard J. Birchall.................... RCAF
C. H. Dickens............................. RAF
James F. Edwards........................ RAF
Joseph C. McCarthy .... RAF

COLOMBIA

Gerhard Thyben

CUBA

Rafael del Pino .... FAR

CZECH REPUBLIC

M. A. Liskutin............................ RAF
Vladimir Remek .... CAF

FINLAND

Eino I. Juutilainen .... FAF
Olli P. Puhakka............................ FAF
Hans H. Wind.............................. FAF

FRANCE

Marcel Albert .... FAF
Valérie André .... FAF
Jacqueline Auriol .... Civilian
Pierre Clostermann.................... RAF
Roland de la Poype .... SAF

GERMANY

Hansgeorg Bätcher .... Luftwaffe
Gottfried Dulias .... Luftwaffe
Adolf Galland...................... Luftwaffe
Hajo Herrmann .... Luftwaffe
Dietrich Hrabak .... Luftwaffe
Walter Krupinski .... Luftwaffe
Kurt Kuhlmey...................... Luftwaffe
Fritz Losigkeit..................... Luftwaffe
Diether Lukesch .... Luftwaffe
Rudolph Opitz .... Luftwaffe
Dietrich Peltz .... Luftwaffe
Günther Rall .... Luftwaffe
Wolfgang Schenck............... Luftwaffe
Heinrich Schmetz................ Luftwaffe
Wolfgang Späte .... Luftwaffe
Ulrich Steinhilper .... Luftwaffe

HONDURAS

Fernando Soto Henriquez........... FAH

HUNGARY

Daniel Holeczy......................... RHAF

ICELAND

Thorsteinn E. Jonsson .... RAF

ISRAEL

Giora Even ....IAF
Louis Lenart ....IAF
Amir Nachumi ....IAF
Arieh Oz ....IAF
Joshua Shani................................IAF

JAPAN

Saburo Sakai................................. IJN

NEPAL

Madan Khatri Chetri..................RNA

NETHERLANDS

Dirk L. Asjes .... RNLAF
HRH Prince Bernhard........... RNLAF
Bram van der Stok .... RAF

NEW ZEALAND

Warren E. Schrader .... RNZAF

NORTH KOREA/UNITED STATES

Kenneth Rowe......................... NKAF

POLAND

Boleslaw Gladych .... PLAF
Miroslaw Hermaszewski .... PLAF
Waclaw Makowski..................... PLAF
Witold Urbanowicz .... PLAF

RUSSIA

Georgii T. Beregovoi .... SAF
Evgeny G. Pepelyaev.................... SAF
Vitali I. Popkov .... SAF
Alexander P. Silantiev .... SAF
Pavel A. Taran .... SAF
Konstantin M. Treshchov............. SAF
Aleksandr Vybornov .... SAF

SPAIN

Jose Larios.................................. SAF

SWEDEN

Olle Klinker .... Civilian

SWITZERLAND

Bertrand Piccard..................... Civilian
Yves Rossy................................... SAF

TURKEY

Sabiha Gökçen .... Civilian

UNITED KINGDOM

A. W. Bedford .... Civilian
Roland P. Beamont .... RAF G.
Leonard Cheshire .... RAF
John Cunningham .... RAF
W. Dennis David .... RAF
Ron Dick...................................... RAF
A.G. Dudgeon............................. RAF
Raynham Hanna.......................... RAF
Sir Patrick Hine .... RAF
F. Desmond Hughes .... RAF
James E. Johnson .... RAF
William Reid................................ RAF
Paul C. Webb .... RAF
Andrew D. Auld .... RN
Eric M. Brown.............................. RN
L. Peter Twiss .... RN
Sir John W. Hackett ....British Army

UNITED STATES

C. Alfred Anderson ....Civilian
Maxie L. Anderson.................Civilian
Brian Binnie...........................Civilian
Connie Bowlin ....Civilian
Richard T. Brickert ....Civilian
John P. Capellupo ....Civilian
Philip M. Condit.....................Civilian
Oliver R. Crawford.................Civilian
E. Patrick Epps.......................Civilian
David L. Ferguson ....Civilian
Linda Finch............................Civilian
David F. Fowler ....Civilian
Richard H. Frost.....................Civilian
Ernest K. Gann ....Civilian
Robert J. Gilliland ....Civilian
James G. Haizlip ....Civilian
Mary Haizlip...........................Civilian
Alfred C. Haynes....................Civilian
Homer H. Hickam, Jr..............Civilian
Bruce J. Hinds ....Civilian
Robert P. Hudgens..................Civilian
Alvin M. Johnston ....Civilian
Anthony W. LeVier ....Civilian
Alan R. Mulally......................Civilian
Thomas P. Poberezny..............Civilian
Robert M. Robbins ....Civilian
Richard G. Rutan ....Civilian
Michael M. Sears....................Civilian
John F. Smith ....Civilian
Evelyn Trout...........................Civilian
Sean D. Tucker ....Civilian
Kermit A. Weeks ....Civilian
Jessie Woods...........................Civilian
Jeana L. Yeager ....Civilian

US AIR FORCE AND US ARMY AIR FORCES

Harry C. Aderholt .... USAF
Paul W. Airey .... USAAF
John R. Alison .... USAAF
Clarence E. Anderson .... USAAF
Lee A. Archer, Jr. .... USAAF
Fred J. Ascani .... USAAF
Regina C. Aune .... USAF
William T. Badham ....Air Service
Rex Barber .... USAAF
Alcide S. Benini......................... USAF
Robert D. Berg ....USAFR
Fredrick C. Blesse..................... USAF
Winton M. Blount................. USAAF
Charles R. Bond, Jr. ....GHQ AF
Charles G. Boyd .... USAF
Devol Brett.............................. USAF
James F. Brewer .... USAAF
A. Raymond Brooks...........Air Service
James Brooks .... USAAF
Harry W. Brown .... USAAF
Roscoe C. Brown, Jr. .... USAAF
William Earl Brown, Jr. .... USAF
Jacksel M. Broughton .... USAF
Douglas Campbell ....Air Service
William A. Campbell............. USAAF
Robert L. Cardenas.................. USAF

John A. Carey .... USAAF
John T. Carney .... USAF
Herbert E. Carter .... USAAF
Richard C. Catledge .... USAF
Charles G. Cleveland .... USAF
Richard E. Cole .... USAAF
Eileen M. Collins .... USAF
Kenneth S. Collins, Sr. .... USAF
Jerry D. Collinsworth .... USAAF
Ramón Colón-López .... USAF
Violet Cowden .... WASP
Barry F. Crawford, Jr. .... USAF
Laurence C. Craigie .... USAAF
Iris Cummings Critchell .... WASP
Woodrow W. Crockett .... USAAF
Kenneth H. Dahlberg .... USAAF
Benjamin O. Davis, Jr. .... USAAF
George E. Day .... USAF
Eugene P. Deatrick .... USAF
Charles B. DeBellevue .... USAF
Joseph R. Deslauriers .... USAF
Israel Del Toro .... USAF
Fredric M. Donohue .... USAF
James H. Doolittle .... USAAF
Clyde B. East .... USAAF
Billy G. Edens .... USAAF
Leon F. Ellis .... USAF
Herman E. Ernst .... USAAF
Frank K. Everest, Jr. .... USAF
Scott Fales .... USAF
Gregory Feest .... USAF
Thomas W. Ferebee .... USAAF
Harold E. Fischer .... USAF
Bernard F. Fisher .... USAF
Wayne L. Fisk .... USAF
James P. Fleming .... USAF
Maxine E. Flournoy .... WASP
Ronald R. Fogleman .... USAF
Hank P. Fowler, Jr. .... USAF
Arnie L. Franklin .... USAF
Carl S. Fraser .... USAF
Fitzhugh L. Fulton, Jr. .... USAF
Francis S. Gabreski .... USAF
Vermont Garrison .... USAF
Ralph D. Gibson .... USAF
Kevin A. Gilroy .... USAF
Robert J. Goebel .... USAAF
Leon José González .... USAAF
Gordon M. Graham .... USAAF
Leo R. Gray .... USAF
Herschel H. Green .... USAAF
Thomas C. Griffin .... USAAF
Thomas E. Griffith, Jr. .... USAF
Robert Gutierrez, Jr. .... USAF
Gail S. Halvorsen .... USAF
Phil Handley .... USAF
Haywood S. Hansell, Jr. .... Air Corps
William S. Harrell .... USAAF
Harold R. Harris .... Air Service
James H. Harvey III .... USAF
Michael V. Hayden .... USAF
Bernice Falk Haydu .... WASP
Thomas L. Hayes .... USAAF
Gloria W. Heath .... WASP
John P. Henebry .... USAAF
David L. Hill .... AVG
James L. Hobson, Jr. .... USAF
Bruce K. Holloway .... USAAF
Robert A. Hoover .... USAAF
Charles A. Horner .... USAF
Edwin W. Horton, Jr. .... USAAF
James H. Howard .... USAAF
William G. Hudson .... USAF
Clayton M. Isaacson .... USAF
Joe M. Jackson .... USAF
Alexander Jefferson .... USAAF
Eldon W. Joersz .... USAF
Gerald W. Johnson .... USAF
James K. Johnson .... USAF
Leon W. Johnson .... USAAF
Paul T. Johnson .... USAF
Robert S. Johnson .... USAAF
Samuel R. Johnson .... USAF
"Scrappy" C. Johnson .... USAF
David M. Jones .... USAAF P
earl Judd .... WASP
James H. Kasler .... USAF
James R. Kirk .... USAAF
William L. Kirk .... USAF
Joseph W. Kittinger, Jr. .... USAF
Robert D. Knapp .... Air Service
Richard A. Knobloch .... USAAF
Walter Kross .... USAF
Michael I. Lampe .... USAF
William R. Lawley .... USAAF
Jerry L. Leatherman .... USAF
Eugene B. LeBailly .... USAF
Curtis E. LeMay .... GHQ AF
Roger C. Locher .... USAF
John S. Loisel .... USAAF
Barbara E. London .... WAFS
Donald S. Lopez .... USAAF
Charles H. MacDonald .... USAAF
Walker M. Mahurin .... USAAF

LeRoy J. Manor .... USAF
Forrest S. McCartney................ USAF
James H. McClain .... USAAF
Carroll W. McColpin .... RAF
Jim M. McCoy .... USAF
Charles E. McGee .... USAF
Melvin A. McKenzie .... USAAF
William T. Meredith................. USAF
Alfred Metz.............................. USAF
Theodore R. Milton .... USAAF
John W. Mitchell.................. USAAF
Joseph H. Moore .... USAAF
Thomas S. Moorman, Jr. .... USAF
Robert K. Morgan .... USAAF
James P. Muri .... USAAF
Frank Murray ....CIA/USAF
Stanton R. Musser.................... USAF
Lloyd W. Newton .... USAF
Joe M. O'Grady........................ USAF
Charles H. Older........................AVG
Robin Olds.............................. USAF
Dolphin D. Overton III .... USAF
John R. Pardo .... USAF
Ralph S. Parr .... USAF
Deanie Parrish..........................WASP
Gaillard R. Peck, Jr. .... USAF
Chesley G. Peterson.................... RAF
Spiro N. Pisanos .... USAAF
Bryce Poe II .... USAF
Henry A. Potter..................... USAAF
Ramsay D. Potts .... USAAF
Cecil W. Powell .... USAF
Elwood R. Quesada .... USAAF
Michael J. Quirk.................... USAAF
Edward F. Rector................... USAAF
Zachary J. Rhyner..................... USAF
Margaret J. Ringenberg.............WASP
Robinson Risner....................... USAF
Richard S. Ritchie..................... USAF
Cesar A. Rodriguez, Jr. .... USAF
Robert Rosenthal.................. USAAF
Ed J. Saylor .... USAF
Russell E. Schleeh .... USAF
Bernard A. Schriever .... USAF
Robert L. Scott, Jr. .... USAAF
Dawn R.B. Seymour .... USAF
Joseph L. Shannon .... AL ANG
Robert J. Shoens .... USAAF
Brian Shul .... USAF
Joseph Smith ....Air Corps
Maurice R. Smith ....Air Service
William Y. Smith .... USAF
William W. Spruance .... USAAF
Harry T. Stewart, Jr. .... USAF
Walter T. Stewart.................. USAAF
Donald J. Strait.....................NJ ANG
Raymond Strasburger .... USAF
Elizabeth B. Strohfus................WASP
Dora D. Strother ....WASP
Leo K. Thorsness...................... USAF
Paul W. Tibbets .... USAAF
Nancy Hopkins Tier .... CAP
Reade Tilley .... RAF
Mark W. Tillman...................... USAF
Robert F. Titus .... USAF
Raymond F. Toliver................... USAF
Dennis Trayno .... USAF
Theodore Van Kirk .... USAAF
George A. Vaughn, Jr. ....Air Service
John J. Voll .... USAAF
Leigh Wade ....Air Service
Henry E. Warden .... USAF
John A. Warden III .... USAF
Wesley Wells .... USAAF
James C. Warren....................... USAF
Florene Miller Watson.............WASP
Robert M. White...................... USAF
John F. Whiteley...............Air Service
Timothy A. Wilkinson.............. USAF
David M. Williams .... USAAF
Robert H. Wolff .... USAAF
Maggie H. Woodward .... USAF
Robert G. Wright, Jr. .... USAF
Charles E. Yeager .... USAAF
Gerald O. Young .... USAF
Hubert Zemke....................... USAAF

US ARMY

John C. Bahnsen, Jr..................... USA
Patrick H. Brady......................... USA
Richard A. Cody......................... USA
David F. Cooper .... USA
Rhonda L. Cornum .... USA
Bruce Crandall .... USA
Michael J. Durant....................... USA
Frederick E. Ferguson................. USA
Ed Freeman............................... USA
Clyde A. Hennies .... USA
Randolph Jones .... USA
Karl H. Maier............................ USA
Lance McElhiney .... USA
Lawrence E. McKay .... USA
Hugh L. Mills, Jr........................ USA
Harold G. Moore........................ USA

Michael J. Novosel, Jr. .... USA
John H. Oswalt .... USA
Ellis D. Parker .... USA
Ralph Puckett, Jr. .... USA
Mathew B. Ridgway .... USA
Edward D. Shames .... USA
Stanley L. Wood .... USA

US MARINE CORPS

James F. Amos .... USMC
Charles F. Bolden, Jr. .... USMC
John F. Bolt .... USMC
Gregory Boyington .... USMC
Marion E. Carl .... USMC
Dean Caswell .... USMC
Jack R. Cram .... USMC
Jefferson J. DeBlanc .... USMC
Archie G. Donahue .... USMC
Joseph J. Foss .... USMC
Robert E. Galer .... USMC
Guy Hunter .... USMC
Thomas H. Miller .... USMC
Frank E. Petersen, Jr. .... USMC
Theodore A. Petras .... USMC
Manfred A. Rietsch .... USMC
James E. Swett .... USMC
Kenneth A. Walsh .... USMC

US NAVY

Richard H. Best .... USN
George H. W. Bush .... USN
Guy P. Bordelon, Jr. .... USN
Cook Cleland .... USN
Randall H. Cunningham .... USN
Dieter Dengler .... USN
Jeremiah A. Denton, Jr. .... USN
William P. Driscoll .... USN
Donald D. Engen .... USN
Kenny Fields .... USN
James H. Flatley, III .... USN
Leon Frankel .... USN
George Gay .... USN
Paul T. Gillcrist .... USN
Thomas J. Hudner, Jr. .... USN
David McCampbell .... USN
Scott McCuskey .... USN
Thomas H. Moorer .... USN
Thomas R. Norris .... USN
C. Joseph Rosbert .... AVG
John Richard Rossi .... AVG
Leonard B. Smith .... USN
James B. Stockdale .... USN
Stanley W. Vejtasa .... USN
Roy M. Voris .... USN
Alexander Vraciu .... USN

NATIONAL AERONAUTICS AND SPACE ADMINISTRATION

Neil A. Armstrong .... NASA
Marta Bohn-Meyer .... NASA
Frank Borman .... NASA
Eugene A. Cernan .... NASA
Charles Conrad, Jr. .... NASA
Richard O. Covey .... NASA
Robert L. Crippen .... NASA
A. Scott Crossfield .... NASA
Charles M. Duke, Jr. .... NASA
Benjamin A. Drew, Jr. .... NASA
Joe H. Engle .... NASA
Robert L. Gibson .... NASA
John H. Glenn, Jr. .... NASA
Susan J. Helms .... NASA
William J. Knight .... NASA
Eugene F. Kranz .... NASA
Walter M. Schirra .... NASA
Brewster H. Shaw, Jr. .... NASA
Alan B. Shepard, Jr. .... NASA
Donald K. Slayton .... NASA
Robert L. Stewart .... NASA

GATHERING OF EAGLES FOUNDATION

Made in the USA
Las Vegas, NV
03 November 2026

57424184R00112